The Administration of Sickness

Also by William Gallois

TIME, RELIGION AND HISTORY
ZOLA: The History of Capitalism

The Administration of Sickness

Medicine and Ethics in Nineteenth-Century Algeria

William Gallois

Softcover reprint of the hardcover 1st edition 2008 978-0-230-50043-3

First published 2008 by
PALGRAVE MACMILLAN

Palgrave Macmillan in the UK is an imprint of Macmillan Publishers Limited, registered in England, company number 785998, of Houndmills, Basingstoke, Hampshire RG21 6XS.

Palgrave Macmillan in the US is a division of St Martin's Press LLC, 175 Fifth Avenue, New York, NY 10010.

Palgrave Macmillan is the global academic imprint of the above companies and has companies and representatives throughout the world.

Palgrave® and Macmillan® are registered trademarks in the United States, the United Kingdom, Europe and other countries.

ISBN 978-1-349-35262-3   ISBN 978-0-230-58260-6 (eBook)
DOI 10.1057/9780230582606

This book is printed on paper suitable for recycling and made from fully managed and sustained forest sources. Logging, pulping and manufacturing processes are expected to conform to the environmental regulations of the country of origin.

A catalogue record for this book is available from the British Library.

Library of Congress Cataloging-in-Publication Data
Gallois, William, 1971–
    The administration of sickness : medicine and ethics in nineteenth-
    century Algeria / William Gallois.
        p. cm.
    Includes bibliographical references and index.

        1. Public health administration – Algeria – History – 19th century.
    2. Medical policy – Algeria – History – 19th century. 3. Medical care – Algeria –
    History – 19th century. 4. Imperialism – Health aspects – Algeria – History –
    19th century. 5. Medical ethics – Algeria – History – 19th century. I. Title.
    [DNLM: 1. Health Policy – history – Algeria. 2. Health Policy – history – France.
    3. Colonialism – history – Algeria. 4. Colonialism – history – France. 5. Delivery
    of Health Care – organization & administration – Algeria. 6. Delivery of Health
    Care – organization & administration – France. 7. Ethics, Medical – history –
    Algeria. 8. Ethics, Medical – history – France. 9. History, 19th Century –
    Algeria. 10. History, 19th Century – France. 11. History, 20th Century – Algeria.
    12. History, 20th Century – France. WA 11 HA4 G173a 2008]
    RA552.A4G35 2008
    362.10965—dc22   2008016413

10 9 8 7 6 5 4 3 2 1
17 16 15 14 13 12 11 10 09 08

Transferred to Digital Printing in 2014

Contents

Acknowledgements

The research for this book was generously supported by the American University of Sharjah, the Mellon Foundation, the School of Oriental and African Studies and Roehampton University. I also benefited from the advice and expertise of a number of history of medicine seminars and conferences at which I presented parts of my project, particularly those run by Mark Jackson at Exeter, Hilary Marwick and Colin Jones at Warwick, the Society for the Social History of Medicine, and Gino Raymond's seminar at Bristol. My colleagues at Roehampton – most especially Sara Pennell, Cornelie Usborne, John Tosh and Meg Arnot – have been great supporters of this project, as were Gez Hawting, David Arnold, Peter Robb, Michael Brett and Tom Tomlinson at SOAS; Bob Cook, Larry Woods and Andy Barnett at AUS; and Daniel Pick at Queen Mary. I should also like to thank the staff at the following libraries and archives: CAOM in Aix-en-Provence, the Bibliothèque Nationale in Algiers, the national archives in Algiers, the Bibliothèque Nationale in Paris and the university libraries of SOAS, Exeter, and the University of London, as well as the London Library, the Women's Library and the British Library. I owe a great debt to the anonymous reader of this book, whose incisive criticisms of the text at different stages of its life greatly improved its argument. I also received great help, encouragement and friendship whilst writing the book from Oliver Craske, Jon Cheetham, Niall O'Flaherty, Ruth Hall, Rebecca Fox, Jake Osborne, Rebecca Niblock, Barbara Wallwork, Norman Wallwork, Ramues Gallois, Suzuka Gallois, Richard Gallois, Sarah Gallois, Edward Gallois, Philip Taylor, Encarni Corcoles, Martin Macmillan, Claire Macmillan, Nick Riley, Sole Riley and Otis Kempinksi.

1
Introduction

1.1 Beginning as it ended

On 5 April 1921, the mayor of Cherchell wrote to his Prefect to complain that migrants from the interior posed a serious threat to the health of his coastal town. 'Long queues of people from Orléanville and Ténès' had tried to enter Cherchell, and when they had been expelled, they had simply tried to come into the town through another gate'.[1] They 'brought with them the threat of typhus...to a settlement which had until then been spared, and the only means we had of defending ourselves against this invasion was to guard all the gates into the town'.[2]

That the Prefect took such problems seriously was made plain when, on 4 May 1921, he wrote to the mayor of Algiers of 'the dangers posed to public health by the large numbers of indigènes who arrived in the prefecture each day, begging for our assistance'.[3] As a means of arresting such population flows and protecting the 'hygiène publique' of the coastal littoral, the Prefect suggested that a form of ticketing system might be introduced so that only those who were shown to be free from typhus could travel towards the capital.

The Prefect knew that such a suggestion would appeal to the mayor of Algiers, for he had traditionally objected to paying for the cost of charitable assistance from the city's budget and had demanded that such expenses be shared by the colonial government and the department of Algiers. Those who were destined to received such gifts were, after all, 'beggars who needed to be purged from Algiers, for they had come to the city not to work or to contribute to the "vie communale", but to exploit the offerings of public charity'.[4]

These local administrative responses to epidemics in the 1920s come after the period 1830–1900 described in this book, but I open my argument with them as examples of a culture of health in Algeria which I suggest began in the nineteenth century and persisted right across the colonial period. When we look, for instance, at the famines and epidemics of the late 1830s, or the period 1867–72, we find that French reactions to such events

1

were framed in more or less precisely the language we find here in the second decade of the twentieth century.

This book, therefore, looks closely at cultures of medicine from nineteenth-century Algeria to discover how permanent, structuring realities were formed in the early decades of the colony and how such cultures of health impacted on both local populations and settler communities. Yet we also need to see that the end reached at Cherchell was arrived at in a particular way, for cultures of health changed over time, so this book offers an account of nineteenth-century Algeria which draws together analyses of both deep-lying structural realities in its history of health and alterations in medical policies, ideas and culture. From the inception of the colony in the 1830s to its fall in 1962, French and Algerian writers consistently saw the French imperial project in the Maghreb as an attempt to *medicalise* Algerian society. This idea of medicalisation lay, I suggest, at the heart of the French attempt to make an Algerian nation in the nineteenth century, its failure to do so and in the encouragement of distinct modes of resistance to French rule.

One of the greatest shifts which took place between the arrival of the French in 1830 and the brutal realities of the world of Cherchell and Algiers in the 1920s was the slow death of the idea of medical imperialism, which invigorated and formed a key part of the early French colonial enterprise, but which can be seen to have collapsed in the examples set out above. That idea of the civilising potential of medicine, and the manner in which disillusionment set in with regard to the idea, amongst local, settler and metropolitan constituencies, runs across the chapters of this book.

In continuing my initial theme of endings and beginnings, let us now look at quotations from the inception and last days of the colony side by side. The first comes from a book written by the politician Pellissier de Raynaud in 1837, the second from Frantz Fanon's resignation note from the Blida-Joinville hospital, which he left in 1953 prior to joining the FLN:

> There exists a dangerous illusion that Algeria presents us with a land ripe for exploitation, a treasure trove into which we can dip our fingers, which needs to be promptly destroyed. Instead we need to see that Africa presents us with the possibility of unleashing the germ of a new people, who might create a gloriously productive new society. It is to this end which I wish to bring discussions, but before doing so we must understand the theatre of operations in which we propose to operate.[5]
>
> If psychiatry is the medical technique that aims to enable man no longer to be a stranger to his environment, I owe it to myself to affirm that the Arab, permanently an alien in his own country, lives in a state of absolute depersonalization.[6]

What I think we see here are expressions of the importance of the idea of medicine in the nineteenth century and, in the twentieth century, the manner in which critics of empire came to understand that the idea of medicine and its implementation constituted one of the most pernicious effects of French colonialism. From Fanon's perspective it is of course quite clear that Pellisier de Raynaud's medical metaphors – 'the germ of a new people' and so on – formed precisely the kind of 'illusion' which Pellisier himself feared, yet a distinctive feature of the function of power relations in colonial settings was that illusions really could become forms of reality, writ large on the *tabula rasa* of the 'theatre of operations on which we propose to operate'. The description of quite how Algerians became 'strangers' in their own 'environment', and not the 'new people' of 'a gloriously productive new society' will be a key task in this book's analysis of the ideas and realities of colonial medicine.

1.2 Scope

This book explores the role which cultures of medicine played in the creation of Algeria in the nineteenth century. Its method is ethical in that it asks whether French medicine was good and just, and whether the medical apparatus of the colonial state can be shown to have improved or diminished the lives of those who became Algerians: both the indigenous peoples of North Africa and European settlers who migrated to the Maghreb.

It identifies three distinct phases in the medical-ethical history of nineteenth-century Algeria. In the first of these stages, lasting from the conquest in 1830 until the 1850s, the *idea* of medicine and of its efficacy was seen to be a crucial form of justification for France's presence in North Africa. A grand plan to medicalise the colony was viewed as being both moral and pragmatic, in that it served the purposes of proving the beneficence of France's civilising occupation and it helped to pacify a hostile and poorly understood environment.

At a second, overlapping, moment through the 1840s '50s and '60s, parts of the grand medical plan were actualised, but it quickly became clear that France could not afford to establish the comprehensive and universal health service that had initially been imagined. The reality of medical provision was that its organisation mirrored the fragmentary character of the colonial state, which was divided between military and civil authorities which tended towards competition rather than collaboration. Medical authorities and conscientious doctors in the capital could look at much of their own work and believe that the imagined medicalised state was coming into being, but the reality for most of the country was that medical care was sporadic, often costly and, in some cases, resisted by locals.

Finally, this situation worsened markedly in the past three decades of the century: racial ideas definitively trumped those of the advocates of universal care, medical provision became increasingly competitive and privatised, plans to train Algerians as doctors and to integrate them into French medical culture collapsed and the grand plan for Algerian medicine was tacitly abandoned. The organisation of medicine in Algeria increasingly mirrored the hostilities between the three key communities which had been created in the nineteenth century: the French army in Algeria, the colons and Algerians.

Much of the broad structure of this narrative of hubristic fantasy and the realities of the poorly managed retrenchment in medicine are well known through the works of Turin, Lorcin, Ageron and Rey-Goldzeiguer, though perhaps less so to Anglophone audiences. What is perhaps less well understood, and where this work marks an original intervention in the field, is the character of the relationship between medicine and ethics in an environment where moral questions were arguably key determinants of shifts in the plan to medicalise the colony. This is also true of the broader field of the history of medicine, whose interest in ethics has tended to focus on rather narrow studies of ethical codes, without examining the ways in which those moral strictures played themselves out in medical practice. A concentration on ethics takes this book towards subjects – such as ideas about the value of life, massacres, pain, torture, war and famine – that have not always been seen as the province of the history of medicine.

I show how a moral culture formed around questions of living, dying, health, sickness, the human and the inhuman, and quite what the consequences of that culture were for colons and indigènes. The chief originality of the book lies, therefore, in this stress on ethics, culture and health, but I hope that each of its chapters adds to existing literatures in distinct ways (in some cases connecting what have traditionally been discrete debates in distinct spheres through the Algerian example).

My title – *The Administration of Sickness* – serves a double purpose, for it suggests first that French imperialism in Algeria brought negative consequences for the land's inhabitants, even when it might have thought itself to be doing good, and, second, it highlights the book's concentration on the administrative structures of the French medical establishment in Algeria. Before the arrival of the French in 1830, there was of course no Algeria, and what I shall contend is that Algeria was made by the French as a sick state over the course of the nineteenth century; as the doctor and novelist Malika Mokeddem intimated when she described Algeria's 'longue fièvre française'.[7]

The second half of this chapter explores the interconnections of history, ethics and health, following Cooter's call for the beginning of a history of medical ethics which looks at the period between Hippocrates and the rise of bioethics in the second half of the twentieth century. Before then, I shall return to Cherchell to explore why the administrative responses cited above are so emblematic of a culture of health in the colony, before looking at the way in which a narrative complex about Algeria built up in the nineteenth century, and at the distinctive contribution of this work in the context of existing work on the history of Algeria and the history of medicine.

Using new sources, the second chapter describes both the contours of the idea of medical imperialism and the practice of medicine in the first four decades of the colony, investigating the manner in which attempts were made to translate 'the idea' into a set of practices. Chapter 3 then looks at how structural forms of dehumanisation were embedded within the humanitarian culture which formed a key part of the moral edifice that structured 'the idea of medicine'.

The heart of the book is found in four chapters which look at the consequences of French medicine for Algerians and at local responses to the project of medicalisation. Chapter 4 studies a massacre at Dahra as an example of how the history of health, and its ethical ramifications, extend beyond the subject matter of the history of medicine, whilst Chapter 5 looks at the manner in which the political and social crises of the period 1867–72 should be viewed in terms of their epidemiological, demographic and ethical consequences for the peoples of Algeria. Both these chapters seek to make clear connections and causal links between, on the one hand, colonial politics, administration and medicine and, on the other, the lives of Algerians. Chapters 6 and 7 then offer the first published accounts of the lives of Algerians working in French medical structures in the period 1870–1900. The moral character of their testimonies of disillusionment and injustice serves a broader argument which runs across the text, beginning with earlier chapters' considerations of Algerian voices such as that of Hamdan Khodja.

Readers familiar with Francophone literatures on the history of medicine in nineteenth-century Algeria and debates on the history of medical ethics may want to concentrate their attention on Chapters 4 to 7. The importance of the ethical claims of those later chapters may, however, not be apparent to readers without specialist knowledge of the field unless they gain an understanding of the development of an idea of medicine and its instantiation described in the first half of this book.

1.3 A return to Cherchell

Aspects of the history of colonial medicine in Algeria have been discussed in the work of authors such as Turin and Lorcin, but one of the key differences of this book is the manner in which it focuses on the ethics of medical encounters and cultures of health. In the examples with which this chapter began, we saw that thinking about medical questions extended well beyond the professional realm of doctors, and that the language and ideas which were used to express judgements about health tended to be deeply moralistic. It was often the ethical conceptualisations of health of soldiers and civil administrators which were as formative of medical cultures as the practice of doctors, so it may be more realistic to see ethics here as a dialogue both amongst colonial groupings and between colonists and the *indigènes*. It was also the case that medical professionals played a large role in describing the sickness of the Algerian body politic outside of their clinical practice.

One of the most striking aspects of the evidence from Cherchell and Algiers cited above is that which is absent from this writing, for what we see here are a group of officials who had no sense of duty towards the sickness of their Algerian subjects. Algerians were, rather, a group to keep segregated from the healthy society of the European cities, and the very act of their being sick could serve as a means by which to judge them, for their only possible contribution to the 'vie communale' was to serve as vectors of disease and destruction. It was understandable, therefore, within the logic of this ethical system, quite why the notion of 'the public' needed to be strictly delimited and why physical obstructions needed to be erected to complement those mental barriers which had established utter human differences between these racial communities.

Another lacuna which runs across these remarks relates to the question as to how Algerians had come to find themselves in such a state in the 1920s. One of the clichés of French colonial medical writing was its insistence that Muslims saw disease as a form of fate which could not therefore be combated, yet French writers were actually more likely to borrow this trope than they were to find it in Algerian society. In these documents we see an assumption that Algerians were be riddled with disease, with no interest displayed in the manner in which such illness spread or, most critically, the manner in which French administrative, political and medical policies had impoverished and enfeebled Algerian populations. There was, for instance, no attempt to consider the social and economic motivations behind such migrations or 'hunger marches' to towns such as Cherchell and Algiers, for the concern of colonial authorities for those within their 'circle of care' was also expressive of a willed blindness to the underlying realities and consequences of the establishment of an Algerian national space. We might also note the presence of another structural feature of this peculiar medico-ethical realm in the form of the spatialisation of a duty of care, in which migrants and other 'outsiders' were seen to be particularly undeserving of assistance.

At best, these subjects could expect to be the recipients of French charity, but a darker side of French moral thinking about health was also at play here, for we read of the need to 'purge' the cities of the undeserving sick, and across this book I shall seek to explore the intimacy with which beneficent ideas of charitable duty lay alongside exterminatory desires within the very particular culture of Algerian health.

This chapter moves on to look at questions of ethics, history and health in detail, but at this early stage it may be useful to explore further the ethical focus of this project. That concentration is driven primarily by texts and voices from the nineteenth century, for it is quite evident that each of the constituencies involved in this story – colons, indigènes, soldiers, politicians, doctors – described their actions in moral terms. What is more, such discussions invariably took place in the public realm and were often published, so there is, for instance, perhaps some sense of surprise for contemporary audiences that moral debates about, and justifications for, exterminatory massacres were a distinct feature of colonial literatures. It is important to stress that this book is not, therefore, primarily a history based on the analysis of secret files and motivations, but an attempt to reconstruct the moral worlds of the communities of colonial Algeria: to describe them in detail, their structures, how they functioned as ideas and as texts, how they changed over time and the effects they had on others.

The medical ethicist Alastair Campbell hints at the importance of such an enterprise when he writes that:

> The facts of colonization and historical intercultural abuse imply that ethics needs to go beyond knowing a culture and respecting its meanings, an area of philosophy prominently championed by Peter Winch (1958). To an interpretative or hermeneutic knowledge of the realities of another culture we need to add an understanding of social disempowerment and marginalization, a strand of thought that is prominent in the work of Foucault and many feminist thinkers.[8]

Where Campbell stresses the manner in which Foucault and feminist thought can afford us a means of seeing ethics in the colonial encounter as a form of dialogue which needs to listen hard for the voice of the dispossessed, one of the advantages of the Algerian situation is that ethical critiques of imperialism which predate the work of thinkers such as Foucault and Fanon can be found in texts by Algerians from across the nineteenth century. In Chapter 2, I introduce the work of Hamdan Khodja, whose work *Le Miroir* appeared in 1833, whilst in Chapters 6 and 7, I suggest that the names of hitherto unknown doctors such as Abdel Kader ben Zahra, Boulouk Bachi and Mohammed ben Saïah be added to the canon of anti-colonial thought.

I use two distinct means of structuring ethical discussion in the book. The first is to begin each chapter with a narrative account of an event which is in some way emblematic of the themes of the chapter, which is then followed with a discussion of a branch of ethics. The second comes in the framing of my arguments around what have come to be seen as the four principles of modern medical ethics: beneficence, non-maleficence, justice and autonomy. Specific chapters concentrate on each of these ideas, but in the act of choosing such frames, I am making a point about the history of medical ethics, for we will see that in medical-ethical literatures it is commonly believed that the practice of healthcare was morally governed almost exclusively by concerns with beneficence and non-maleficence until late modernity. Such beliefs are wrongheaded in a whole series of ways, chief amongst which is the fact that medical encounters in places like nineteenth-century Algeria involved moral confrontations between cultures which mistook certain forms of scientific progress for ethical superiority, failing to recognise that other societies might possess more complex medical-ethical frameworks than those which existed in Europe. In this sense, the French project of medical imperialism was blind to the realities of the culture on which it sought to operate.

1.4   Narratives of health

As well as being a history of health, this is therefore a book about narratives and their effects on people. It studies the role that medical ideas played in the generation of a narrative complex about Algeria in the nineteenth century. There were of course other distinct genres and subjects which played important roles in the Algerian complex – the environment, race and history for example – but it is both the case that medical thinking tended to feature in these other areas of explanation and that medical ideas arguably served as a substrata which underpinned the overall complex. This primacy came partly from medicine's status in French culture at that time and in part because medicine came to be seen as an effective means of expressing the ethical desires which structured French colonialism in Algeria. Medicine was seen to embody what we may see as a kind of continuum in which theory, practice and ethics were all expressed in a very human form in the truth of the medical encounter in which one man healed another. Existing literatures have tended to underestimate the significance of the textual creation of Algeria as well the idea of medicine and its place within a wider narrative of beneficent colonialism.

In one way this is, then, a text about texts, because what this book is doing is using the massive literatures of the nineteenth century, most of which lie undiscussed in current work, to establish how Algeria was made. Now it may be objected that Algeria was not made in texts but I would argue that in two crucial ways this was the case: first, the complex of ideas we find in texts was an incredibly powerful determinant of behaviour and, second, colonial actors used texts as a means of reflecting on the meaning of their actions. In both cases there existed a dominant sense of ethics in the colony, for texts described justifications and rationales, both before and after events, which depended for their strength on moral conceptions of the world.

Using moral language as a means of teasing out the complexities of nineteenth-century colonial culture is not a simple process, in part because the culture which it seeks to describe was by no means a well-ordered place with clear distinctions in thought and policy between rival camps and factions. To take only six debates – the relationship between the civil and military authorities, immigration, the treatment of locals, religion, capitalism and the future of the colony – it was quite clear that the nineteenth-century

Algerian political realm was a malleable and changeable place, in which new alliances and strategies could emerge as time moved on and circumstances changed. Yet within this rather chaotic policy-making environment, it was quite clear that doctors played an absolutely critical role in creating Algeria, in making it safe and in supporting a form of civilisational rhetoric which remained constant for most of the period, even whilst the immediate fabric of political discussion changed. The legitimising power of scientific and medical literatures was sufficiently great for their popularity to extend well beyond the writings of doctors, and across the book, we will see the more general dominance of the medical lens through which Algeria was viewed.

Medicine was also a fact as well as a theory. It was practised on French and Algerian patients, administrative structures were established to provide for systems of healthcare and people's lives were saved or diminished through their contact with French healthcare. This book therefore explores the relationship between ideas of medicine and their practice, and it concludes that nineteenth-century writers like Kob and Le Pays de Bourjolly were right to critique the culture which led to the creation of a set of fantasies about France's role in Algeria, for those dreams were never likely to be matched by the realities of the colonial state.

The relationship between cultural production and medicalisation was of course not rare in this period. Bewell has noted that 'it was common in this era to figure the colonies as sick environments in need of the therapeutic British hand', in which 'place-centred readings of disease become justifications for colonial expansion'.[9] The arts were of course classic means by which such a 'place-centred' approach could stress the essential difference and strangeness of the Algerian environment, making it a place in which very special kinds of European practices could be countenanced; a distinct moral zone to be operated on and which might be treated in ways which could not be imagined in the metropole.

It is therefore of critical importance that in looking at texts by doctors and others that we pay close attention to the moral worlds which were being created in language (both written and pictorial) in the nineteenth century, and it is for this reason that, in Chapter 3, I will be as interested in the art and ideas of Barbara Leigh Bodichon as with the medical practice of her husband Eugène.

A consideration of cultural practice also needs to address the question of what was and was not represented in art, literature and travel writing. The chief absence in such work was of course any sense of the great changes which were being made in Algeria from the perspective of the majority population, for almost no French cultural producers spoke Arabic or other local languages and they did not feel this as any kind of impediment for they were of course almost universally sure of the realities they saw before their eyes and which they discussed in colon circles. A second lack, in general, was any consideration of what were a whole series of blind spots in both creative and social scientific literatures, for while Algerians and Algeria were subjected to relentless social scientific, medical and psychological analysis, areas such as famine and concomitant epidemics and refugee crises lie almost absent from literatures of the nineteenth century.

Where such descriptions did exist, they reflected that moral textualisation of nineteenth-century Algerian society which we now see as telling us rather more about colonial culture than its notional subject matter. To take one example, the work of the nineteenth-century painter Gustave Guillaumet is revealing – in pieces such as *Rue à El-Kantara*, *La Séguia* and *La razzia* – of the manner in which French representations of the Algerian dead can be said to have revealed a kind of moral stumbling, for even when individual bodies lie in the foreground of his pictures, Guillaumet did not feel able to grant Algerian corpses facial features.[10] We will need to look further at that combination of an instinct to dehumanise, forms of repulsion and guilt, which we see in such art, which was also apparent in the notes with which Guillaumet accompanied his representations of famine.[11] That writing presented Algerians as a people who accepted famine and the death of family members as being the will of God, for it was determined to avoid any explanation of such catastrophes which considered France's socio-political role in their formation. When Guillaumet heard local people praying in such circumstances, he believed that 'their prayers were no more than a form of funereal chant'.[12] 'Is this your wife' he asked one 'poor unfortunate', to which he received the reply 'No, my wife is dead; this is my mother, who is also dead', 'And in completing this thought with a gesture', Guillaumet claimed, 'he seemed to say: "This is the Arab's destiny" '.[13] These words – 'il semblait dire: "C'est la destinée de l'Arabe" ' – and all that they express about French hopes, actions and views of Algeria, based as they are on the speculative interpretation of a gesture, are ones which I track across this book, for Guillaumet's assertion that this was the way in which Algerians saw their fate was an important prop in that edifice of ideas which allowed France to abnegate responsibility for its terrible work in the colony.

1.5   Context

The originality of this book lies partly in the new documents which it studies and in the new insights into the Algerian past which are afforded in viewing that past through the history of health and ethics. The book also draws on a series of brilliant existing literatures in the history of Algeria and the history of colonial medicine. Most of my engagement with those texts takes place in the body of the work, but it remains useful to signal both the debt which my argument owes to key texts in the field and to try to explain how this book develops their arguments.

One of the most important writers on the period – Patricia Lorcin – has written that in 'Algeria…the examination of colonial medicine has been limited',[14] contending that this is somewhat disappointing in terms of the broader history of medicine for

> Historical research along these lines has concentrated on the end of the nineteenth century, when the bulk of the European colonies were acquired and medicine reached its apogee in imperial ideology, but Algeria was in the vanguard of the process [in creating the idea of colonial medicine]. This was due, on the one hand, to the large number of physicians that were present in the colony as of its inception and, on the other, in spite of the enormous progress medical science and especially epidemiology was yet to make before the end of the century, to the preeminence of French medicine and science throughout the first half of the century.[15]

The chief exception to this gap is the work of Turin, though this book will also consider work on medicine by Lorcin herself, along with more general studies of Algeria in the nineteenth century by, amongst others, Ageron, Rey-Goldzeiguer, Clancy-Smith, Marcovich, Osborne and Micoleau-Sicault. Turin's work – which looks at medicine alongside religion and education in the period 1830–80 – is of exceptional importance, not only for the manner in which it sets out the character of the French project of medical imperialism, but also for the fashion in which the ideas which underpinned that enterprise are compared with those which governed other aspects of imperialism. It is education, above all, which interested Turin, for her overarching aim was to trace the roots of Algerian resistance to the French, and of the three spheres she looked at, this was most apparent in schools; though she helpfully connects this struggle to the medical difficulties the French encountered in areas such as vaccination programmes. I try to avoid too heavy a reliance on Turin's sources and arguments, though a part of the aim of Chapter 3 is to introduce the work of Turin and other Francophone writers to Anglophone audiences, for they are rarely discussed in comparative studies of colonial medicine.

One of the goals of this book is to extend our knowledge of the history of French medicine in Algeria to the end of the nineteenth century and into the twentieth century, for Turin's work finishes in 1880, and she is not therefore wholly able to capture the important shifts which took place in the early Third Republic. As Lorcin suggests, the Algerian example was of more general significance in the history of colonial medicine, for in some ways it served as a testing ground for ideas and practices which were later exported by France to Tunisia, Morocco, other parts of Africa and to South-East Asia. In works by Gallagher, on Tunisia, and Bidwell, on Morocco, we see that many of the military and medical personnel who imagined medicalised forms of imperialism in those colonies had served in Algeria and that their reconceptualisations of the idea of medicine owed much to their experiences there.

This book is more instinctively critical of French medicine than Turin's work, for I do not accept the binary of modernity and tradition which she uses to characterise the encounter between France and Algeria, though I can see why she believes that the character of 'ordinary' doctors, teachers and priests in Algeria is deserving of some rehabilitation. My objection to this notion is, however, that doctors and teachers were not innocents in the project of colonialism, but often self-consciously saw themselves as key producers of an imperial ideology and culture. The importance of a number of doctors – such as Eugène Bodichon, the Bertherands and Warnier – to the life and coming-into-being of the colony is considered not only by Turin but also by writers such as Emerit and Rey-Goldzeiguer. Where I hope this book extends such discussions is in analysing a much larger field of medical writing from the colony so as to more effectively describe the moral world associated with medicine in Algeria; or rather, the moral conflicts and points of consensus which characterised that world, and the manner in which some norms changed over time whilst others remained constant.

There is a broader point to be made about methodology and the study of Algeria here, for I find it instructive to compare existing work on the colony with Davis's *Late Victorian Holocausts*. Davis's methodological eclecticism – he links studies of genocide with accounts of climate changed and capitalist development – and his determination to see historical evidence as a means to and for the making of ethical claims contrasts with an instinctive conservatism in much historical writing, which is reluctant to espouse moral judgements, for fear both of an abandonment of neutrality and uncertainty regarding the causal influence of distinct spheres upon each other in the cultures of the past. This may seem a slightly unfair judgement on work by writers such as Ageron and Rey-Goldzeiguer, which have hardly been afraid to critique the actions of the French and the colons, but it is my conviction that new accounts of connectedness in Algerian history, which draw on the claims of writers such as Davis, can now be written. Ageron once remarked that his work was not 'histoire engagée...dite...anti-colonialiste',[16] yet I think a case can now be made that precisely such histories can build on his great histories. Ageron's works are models of careful scholarship and of the detailed delineation of shifts of ideas and policies in Algeria. In many senses he created an empirical base upon which all subsequent writers on Algeria have been able to work, and it is for this reason that I operate with the conceit that Ageron's work now allows others to practice a more open 'histoire engagée' dite 'anti-colonialiste'.

In its shift towards ethics, this book aims to become a contribution to the history of health rather than strictly to the history of medicine. By the history of health, I mean to describe a field which has a rather broader optic than traditional studies of medicine, for it concerns itself with cultural, social, political and administrative factors which impinged on the health of a population, as well as medical structures which emerged from those broader spheres. Traditional literatures on the history of colonial medicine in Algeria tended to concentrate on advances made by men such as Alphonse Laveran and the significance of the development of institutions such as the Institut Pasteur in Algiers, though it is also true that the instincts of the social history of medicine, especially in the colonial field, have focused much more on the consequences and limitations of colonial medical structures.

Where my conception of the history of health moves beyond the concerns of both of these literatures is in its determination to express the causality and interdependency of cultural ideas – such as the desirability of exterminating native populations – with both demographic consequences and the relationship of such notions to what may seem to be more benevolent conceptualisations of a civilising medical mission. Put bluntly, the very idea that an Algerian deserved to die depended upon medicalised systems of race, and the exterminatory practices of the French army were in some senses assured by an ethical argument which suggested that massacres flowed as natural consequences from situations where beneficent gifts such as medical progress were spurned.

One of the goals of a history of health must be to study a much wider range of primary sources than those immediately connected with health and medicine, and it is for this reason that I supplement study of doctors such as Bertherand and Warnier, along with lesser known medical figures not considered in existing literatures, with the writing of soldiers such as Létang, artists such as Barbara Leigh Bodichon, legal scholars such as Béquet, politicians such as Coinze d'Altroff and travel writers such as Vereker. Similarly, I look at a number of works written by doctors on non-medical issues, which evince their social and political importance in the colony and the manner in which their status allowed them to have a powerful role in the generation of ideas about Algeria. Part of the aim of this eclecticism is to follow Said's model in *Orientalism* where he sought to show how cultural worlds which existed in the nineteenth century were possessed of a strangeness and completeness which can easily escape us today.

As well as the work of Davis, this work draws more generally on the themes and approaches of the history of colonial medicine. The Marxist, post-colonial, Foucauldian and Subaltern approaches of much work in the field has afforded it an instinctive concern with colonial subjects and the means by which medicine can be viewed as repressive as well as a form of liberation. This sub-field also displays as an instinctive openness towards both the analysis of moral cultures in the past and the use of ethical language in history-writing today.

The Algerian case certainly seems to fit Bewell's contention that 'medicine [lies] at the heart of the politics of colonialism'.[17] Work by figures such as

Arnold, Marks, Watts, Lyons, Farley, Macleod and Lewis has tended to concentrate rather more on the Indian subcontinent and central Africa than on the Maghreb, and I believe that there is a need both to place North Africa into the context of existing debates and to see what the Algerian example can add to such discussions. My claim that medicine can be seen as an emblematic locus of Algerian colonial experience chimes with Mackenzie's neat remark that 'Through medicine and its related disciplines, the West assured itself that it was capable of diagnosing the bodily ills of the indigenous people of empire as part of its wider cultural, political and economic project.'[18]

An additional strength of this field has also been its attendance to the fallacy that an essential difference existed between nineteenth-century Western cultures, which were reputedly fundamentally individualistic, and their non-Western counterparts, which were invariably read as being group-centred. As Worboys and others have noted, such a view wilfully ignored differences of gender, class and race in European culture.[19] As this book unfolds, my argument will also need to engage with more works from this canon of the history of colonial medicine. There is much to be gained from comparative analyses such as Jill Dias's account of the role played by Portuguese imperialism in the spread of disease in Angola and from the methodological debates which have been raised in such works.[20]

A pertinent example of a work in the field which offers important insights into the Algerian situation, in terms both of commonalties and differences, is McLellan's 'Science, Medicine and French Colonialism in Old-Regime Haiti'.[21] His double insight into the manner in which science and medicine were used as tools in an earlier imperial setting in Haiti is that we must acknowledge the complexity of colonial medical-scientific cultures, and how we must see that, although described as 'progressive historical forces', they were as likely to be used to support punitive and regressive forms of politics as they were to be liberatory.[22] The contrast between McLellan's work and others' studies of French medical imperialism in that era – such as Grove's work on Mauritius[23] – is quite marked for his conclusion is that:

> French colonial science at the end of the eighteenth century did not
> directly serve to advance either capitalism or human freedom, so much
> as to further the retrogressive systems of mercantilism and chattel slav-
> ery, systems essentially at odds with the victorious industrial capitalism
> that developed in the nineteenth century.[24]

We shall see that in some ways the Algerian case follows the Haitian template, yet its different moment – coming at the time of 'victorious industrial capitalism' – led to medicine becoming used as a tool of much more diffuse centres of power than was the case in the *Ancien Régime*.

Other aspects of the Algerian experience seem remarkably similar to the culture McLellan describes, for the importance of the conflict which he observed between 'the local medical establishment of Saint Domingue and the powerful medical branch of the royal navy based in France' had parallels in a whole series of power struggles in mid-nineteenth-century Algeria, which were to be of terrible import for the health of the nation.[25] The 'self-governing' character of that local medical establishment and its creation of a medico-moral culture quite distinct from its metropolitan origins also had direct parallels with what I claim to be the special ethical world of health, or worlds of health, in nineteenth-century Algeria.[26]

An additional frame of reference across this book will be the history of medicine in France for, as we have seen, it is quite erroneous to see the metropole and the colony as wholly separate realms. There was, for example, a strong correlation between war, medicine and nation-making in France and Algeria. The struggle of the Franco-Prussian War coincided almost exactly with the climax of the last major rebellions in nineteenth-century Algeria and the attempt to pacify this new national space in North Africa. In both cases, a 'humanitarian' identification of medicine with the state had profound social consequences. As Cooter and Sturdy remark of the metropole:

> Red Cross organizations came to serve as conduits for channelling civilian energies into the conduct of the war, and as a means of linking the pursuit of war to the pursuit of national identity more generally. Under the cover of this ambiguous humanitarianism, medical men were able to create for themselves a leading role in the management of modern warfare. In this instance, wartime medicine contributed to the consolidation of the nation state precisely because it was seen to fulfil a caring and curing role that could ostensibly be distanced from the particularistic national interests that it in fact served.[27]

In Algeria, from the perspective of colonists, there was almost certainly not quite such a need to sugar-coat the 'particularistic national interests' which civilian medicine served in the interests of the state, since a very high level of aggression towards any opponents of the colonial state was increasing at this time, and we may argue that the connection identified by Cooter and Sturdy, and by Taithe in his work on the Red Cross in the Franco-Prussian War, had arguably been developed at an earlier point in Algeria. 'Humanitarian' medical ideas are of special medical interest to this book because their development coincided with what could most neutrally be described as 'demographic collapse' in Algeria, and because they would seem to present a form of medical-ethical archetype, where the presumed beneficence of outcome, which is enshrined in the humanitarian intention, is assumed to hold moral sway over any objections which might be raised on the basis of the actuality of medical and demographic results.

In comparing the colony and metropole, we might also mention the broader role played by medicine in nineteenth-century French modernity, capitalism and political life. As LaBerge and Feingold observe:

> French medical culture reflected dominant themes of contemporary society and politics. The dialectic which ran throughout the century between liberalism and socialism – both legacies of the French Revolution – appeared in medical terms as tension between liberal medicine (private practice) and social, bureaucratic medicine.[28]

This trend was in many ways even more interesting in Algeria than it was in France, for as I have suggested it was in the colony that one found the explosive combination of a variety of medical providers as well as an equally diverse collection of political ideologies and ideologues. This led to the formation of intriguing alliances such as that between Saint-Simonians and the army or between liberal deputies and the interests of Marseille trading enterprises. Algeria was prized as a laboratory in which experiments in modernity could be undertaken at a distance from the metropole, such as the creation of national systems of healthcare in the 1840s, which were not to be introduced for many decades in France. As Léonard remarked, 'Colonisation is, to a certain extent, a form of schooling for metropolitan medicine.'[29]

Finally, the particularities of the development of French capitalism are worth mentioning, for while the term 'capitaliste' was first coined in France – in the last days of the *Ancien Régime* as a means of describing those who exploited land and materials for speculative profit and growth – French capitalism developed in a curious fashion in the first half of the nineteenth century. What some have described as *dirigisme* may be more accurately portrayed as a capitalism of collusion – between interest groups in politics and industry who saw the mutual benefits which might be had if markets were arranged in monopolistic or oligopolistic fashions. An important feature of this particular kind of market-making, which was arguably to not to change significantly until the past 15 years of the century, was the brutal treatment of workers as units of productive capacity, familiar from Marx's descriptions of this time. This first phase of monopoly capitalism was arguably transformed in the final years of the century as philanthropic and state socialist corrections to the market were understood to be both politically expedient and capable of inducing higher levels of long-run productivity.

Where Algeria arguably differed from France in the playing out of this narrative was that the moment when a more paternalist capitalism developed in France coincided with the rise of colon culture in Algeria. Where there had been something of a paternalist, humanitarianism displayed towards Algerians in the period 1830–70, after that point colonists increasingly insisted upon the need for a capitalist culture closer in spirit to earlier French monopoly capitalism. It may reasonably be argued here that the differences I am trying to parse between periods of time are meaningless because, while they may reflect some changes in mentality, they express little of the profound similarity of consequences which Algerian endured in the face of French capitalism in the 1840s and the 1890s: expropriation of their land, their being forced into trading arrangements with local monopolists, where there had previously been an efficient market in goods and services, the changing ecology of Algerian agriculture as French export markets were prized over domestic survival, the eviction of Algerians from their traditional lands into the cities and unfamiliar forms of life and work and the new brutality of the patterns of work imposed on Algerian labourers by French masters. It is also true that at both these times, the lack of regulation of Algerian markets was determined by a mania which lasted for decades regarding the question of whether Algeria would ever 'pay for itself' and become a net contributor to rather than a drain on the national balance sheet.

The formation and consequences of such developments in the colony are studied in much greater detail in Chapter 5, but let us now move from this chapter's earlier foregrounding of the importance of ethics to this project to offer a more detailed account of the history of medical ethics and this book's contributions to the field, beginning with an example of one doctor's work in the colony.

1.6 Parsing Emile Bertherand

Emile Bertherand was one of the longest-serving doctors in the Algerian colony. His writing career began with a valorisation of the idea of medical imperialism, and later included a critique of the manner in which it was implemented, as well as texts on Islam and guides to the practice of healthy living in Algeria. His status as an authoritative voice on medicine and the colony was revealed by the long list of honours he was awarded, which included a directorship of the *Gazette Médicale de l'Algérie*, membership of the Société impériale de Médecine de Constantinople, along with the ranks of Chevalier de la Légion-d'Honneur, Commandeur de l'Ordre royal et militaire du Christ de Portugal, Officier de l'Ordre de Nicham Ifthikhar de Tunis, de l'Ordre du Medjidié de Turquie and Chevalier de l'Ordre de Danebrog.

He was a writer who prided himself on the acuity of the insights which he was able to offer his readership about Algerians, and here I wish to consider three of his works in this regard. In *Médecine et Hygiène des Arabes* (1855) Bertherand made it clear that the central task of medicine was a correction of Algerian culture, 'to ameliorate and modify the erroneous ideas upon which the culture was founded'.[30] This 'vast undertaking of glorious social renovation' called for doctors to undertake an ethical-medical study of the 'anatomie morale' of the Arabs.[31] Yet, as Bertherand was to explain:

> A people – and the Arab people most especially – cannot be known and judged at first glance. Careful research must be undertaken on the principles of their character and their modes of thought, the analysis of the nature of their national fibre, and surveying of their idiosyncratic politics in all its forms. An 'anatomie morale' is hard for a conscientious observer to undertake because his secular method will run up against the blind superstition of his subjects, their belief in the mysteries of life, and their routines which are determined by fatalism.[32]

As was the case with other leading colonial doctors, Bertherand's veneration of medicine as an emblem of secularism and the scientific method was not quite what it seemed. His 'anatomie morale' turned out really to be a form of moral history glossed with a scientific sheen. We might also note both that by this time missionaries had begun to gain influence in the colonial medical state and that the victory he claimed for reason over superstition in France and French medicine was hardly secure in the metropole (leaving aside the 'secular' character of many forms of local medical practices). In fact, one might argue that what was truly evangelical was Bertherand's faith in the capacity of 'Medicine, that powerful lever which allows us to tend to the physical needs of the masses, to wear away their superstitious laziness, to gradually rouse them form their torpor and the stale and static qualities of their Islamic culture.'[33]

Unlike most French writers, Bertherand did have some knowledge of the history of Arab medicine, but such understanding needed to be constrained within a narrative which stressed a strict dichotomy between the golden age of Arab medicine – 'that ancient and glorious era' – and the current impoverishment of Islamic culture and medicine.[34] Bertherand argued that the Qur'anic veneration of doctors and science presented a great opportunity for the French to establish their doctors as the high priests of modernity, which would thus allow medicine to play a redemptive role in healing history as much as in curing individual patients, for the glorious Arab Middle Ages could now be fused to the brilliant European present. Bertherand also admitted that proof of the Arab's desire for a medicalised state could be seen in the hospitals of France's enemy Abdel-Kader, for Bertherand had great admiration for the manner in which they enshrined in legislation the hierarchy of a medical state and assured its people of the value that leaders placed on the provision of good medicine.[35]

While Bertherand's works show that he understood local culture in theory, they also reveal that his understanding of such lives in practice was limited. This is made plain in his study *On Suicide amongst the Muslims of*

Algeria of 1875, which was published in both French and Arabic so that its lessons would be available to both its subjects and their masters. In this book, Bertherand tried to account for the great rise in Algerian suicides which had taken place under colonialism whilst also setting out a series of admonitions against suicide, most which were based on his readings of the teachings of Islam.

Bertherand alleged that the causes of most suicides were trivial reasons, such as jealousy and family disputes.[36] Given that this was the case, he 'called on the consciences of Algerian Muslims' asking why they would choose to kill themselves when 'they had never before lived in a state which offered more protection, benevolence, enlightened justice and forms of assistance than the French colonial government'.[37] A lesson in religion and morality followed, with Bertherand asking why people felt they had to take their own lives, observing that to do so was an act of total selfishness and abandonment of 'what we call Society'.[38] 'In truth', he said, 'those who kill themselves no longer believe in anything: neither their country – for it was a bad citizen who abandoned their post – nor their religion, for as the Qur'an said "man should die only by the will of God." '[39]

What Bertherand revealed here were the limits to medical sociology and moralising as so commonly practised in the colony, for the reality is that his published work was based on no kind of scientific study of the causes of suicide amongst Muslims. In his book, there was no indication that he talked to the families of the dead, nor to the hospitalised who had attempted suicide, for his preferred strategy was to presume that the reasons behind such suicides were sociologically trivial and best remedied through the admonitions of a Frenchman for locals to become better Muslims.

He offered no sense of the psychological reasoning which we find in his near contemporary Durkheim's work *On Suicide* (1897) precisely because, I suspect, Bertherand created this fantastic explanation of self-harm as a means of avoiding a consideration of the *anomie* which Durkheim identified as being so central to the motivation of the suicide. In Bertherand's account the Algerian self was whole but it simply needed to be redeemed, when the reality of the Algerian experience was that both senses of nationhood and religion, which Bertherand claimed were essential components of identity, had been mutilated by the French. What is more, the offer of entry into personhood in terms of becoming French was theoretical rather than practical: for Muslims it involved abandoning their chief pillar of selfhood, for Jews it came to involve making themselves a political target for both Algerian Muslims and the French, and for Arab doctors it involved settling for an adjunct status, institutional humiliation and a career determined by the whims of others.

Bertherand set out Islam's injunctions against suicide, but then failed to ask why people might go against the teachings of their faith in killing themselves. He could not address this question because to do so would have involved removing those spectacles through which Algeria was viewed in the beneficent model of colonial medicine. In concerning himself with ideas and speculation as to why suicides took place, and the manner in which they contradicted what he believed to be the moral codes of the land, Bertherand stood as an exemplar of the fantastic manner in which the medical mission was concerned with ideas over realities, and ethical speculation in place of an analysis of facts. We see this again in Bertherand's work on *Hygiène Musulmane* (1874) in which he directly addressed his Muslim subjects once more, stating that:

> The French authorities have taken every possible opportunity to show you, in her reasoning and in the facts of her action, how close to her heart is the idea of bettering your lives, from the point of view of justice, commerce, agricultural work... of the administration of all your interests.[40]

This was most specifically seen in the arena of health, for:

> The government's is constantly and sincerely trying to fulfil the mission that Providence has granted her in Algeria in the most conscientious manner possible; in its concern for the preciousness of your health and that of your children and families. You will surely have found proof of these sentiments in the institution of free healthcare provision in each of the bureaux arabes.[41]

Not that the bureaux arabes concerned themselves with the numbers of Algerians who killed themselves, those who died in famine, or, worst of all, those who migrated from famines to seek help elsewhere and to die in what were to them foreign lands.

I have chosen to introduce the work of Emile Bertherand as a means of extending this chapter's discussion of health, ethics and history for two reasons. The first is pragmatic since Bertherand worked across most of the period studied in this book, and the consistency of his approach demonstrates the manner in which a particular idea of medicine, and a set of humanitarian structures which logically followed from that idea, structured an environment of health across nineteenth-century Algeria.

The second reason is rather more important in the context of this chapter, for I think the idea of studying Bertherand raises questions as to the difficulty of both attempting to study a moral culture in the past whilst at the same time developing historical arguments framed around ethics in the present. I have called this section 'Parsing Emile Bertherand' because I think there is a methodological need to try to separate out my desire to describe a world in the past and the means by which I will attempt to reconstruct aspects of that world in the present. There is in fact an added difficulty in this enterprise which also needs to be highlighted: namely that the ethically inclined study of medicine and health needs to acknowledge both gaps between texts which sought to codify medical ethics and the realities of medical practice, as well as the need to ensure that connections are made between people's health and an ethical culture which extended well beyond the realm of the writings and codes of medical professionals.

To try to talk meaningfully about how such questions need to be parsed apart and pieced together, this chapter goes on to look separately at the field of the history of medical ethics, the question as to how we might write about ethics historically, the ethical potential of medicine in the nineteenth century and the historical-ethical analysis of medical choices.

It should be clear that my stress on health as well as medicine in this chapter reflects a broader concern with the limits of those fields which have addressed questions such as those posed in this book. I am interested in living and dying in nineteenth-century Algeria, as well as questions of healing, and in this I am following the instincts of the ethics of Islamic medicine which stress questions of the quality of life, health in the context of continuous lives rather than simply moments of medical intervention, and the need to acknowledge broader questions about society and community in talking about health. It is for this reason that Chapter 4 of this book addresses the question as to why massacres should be considered a legitimate area of study in the history of medicine, whilst Chapter 5 seeks to connect questions of health, medicine and epidemiology to war, politics and capitalism in the period 1868–72. In taking such an approach I am also trying to be mindful of Shula Marks's comment that in spite of the great success of the development of a field of the history of colonial medicine, there remains the danger that

> In our recent concern with discourses and texts, we may be in danger of forgetting that there is another history of actual morbidity and mortality, difficult as these may be to determine especially – but not uniquely – in colonial situations, and of actual therapeutic practices and institutions, in all their ambiguities and contradictions.[42]

There are risks involved in the approach I am outlining, but I cannot otherwise see how one might connect a reconstruction of the ethical world of doctors in nineteenth-century Algeria to a study of health, demography and the manner in which Algerians's lives were structured by medical-moral ideas. In a sense, the danger of confusion entailed in my method reflects the muddled character of a nineteenth-century culture where clear lines were not established in theory or practice between questions of medicine, morality and the nascent human sciences. For instance, in Algeria medicine and history were, as we saw in the work of Bertherand, intimately connected since they both served an idea of beneficent colonialism and were fused together in that idea: historical analysis proved the moral and material decline of the Algerian people, whilst medicine assumed a redemptive role in reorienting these people towards a modern historical present. Medicine and history operated in tandem: as forms of practical morality and as the means to ground choices based upon common judgements of the character of Algerian society.

1.7   The history of medical ethics

The history of medical ethics is not a well-developed field. At present it suffers from a number of important lacunae, some of which have been identified by writers such as Cooter, and some of which I will need to outline here. Such gaps also of course present opportunities in imagining and shaping the field, so my chief concern here will be to try to set out what I believe to be four areas in which the history of medical ethics might develop, rather than simply concentrating on its existing limitations.

The first such area relates to the ahistoricism of most writing on medical ethics and what Cooter called 'the resistible rise of the history of medical ethics'. In 1993, Baker argued not only that 'for the best and worst of reasons, there has been little primary research in the history of medical ethics', but also that this

> predilection for ahistoricism is reinforced for the history of medical ethics by the myth of the Hippocratic Footnote, that is, the view that Western medical ethics was established in Greece by the Hippocratic Oath and for the subsequent two millennia amounted to little more than footnotes interpreting this foundational commitment.[43]

Cooter concurred with this verdict of ahistoricism, suggesting that,

> If history is the study of the present at the remove of the past, there have been few signs of that pursuit around medical ethics. The concerns that have come to preoccupy so many philosophers, theologians, feminists, lawyers, novelists, journalists and legislators – not to mention doctors, patients and research scientists – appear to have hardly stirred historians.[44]

This absence of a developed field of the history of medical ethics might therefore be posited as an explanation for ahistorical claims such as Beauchamp and Childress's contention that, 'Medical ethics enjoyed a remarkable degree of continuity from the days of Hippocrates until its long-standing traditions began to be supplanted, or at least supplemented, around the middle of the twentieth century.'[45] This assertion is problematic in all sorts of ways for it displays a number of faults which are common in the field – such as West-centrism and a strict emphasis on ethical codes – but a simple and telling rejoinder which needs to be made to Beauchamp and Childress is that their claim of continuity is based on ignorance and an absence of evidence, rather than any body of scholarship which has actually proven that there was little development in medical-ethical ideas between the Greeks and the modern West.

It may be the case that the medical-ethical cultures of ancient Greece and the modern West at the outset of the Second World War appeared to be rather similar – though one doubts that they resembled each other as much as the authors suggest – but this is not to say that there may not have been many cultures inside and outside the West which had very different ethical cultures in the thousands of years between the ancient and modern worlds. The imaginative paucity of Beauchamp's and Childress's method here is indicative of the broader problem with the obsession with stasis rather than change in historical studies of medical ethics, and the consequent inability of such studies to conceptualise encounters between peoples with differing conceptualisations of ethics, medicine or health. In temporal and methodological senses, this book therefore aims to contribute to the field in presuming change rather than stasis; and in proving that the lost history of medical ethics between Hippocrates and the Second World War was a more complex affair than one would infer from current literatures.

Such a claim also in some senses adds to and in some senses complicates Toulmin's suggestion that 'medicine saved the life of ethics' in the modern West.[46] This position was based upon an analysis of what Toulmin saw as the increasing disconnection between ethical philosophical thought in the nineteenth century and the analysis of lived realities, which changed in the period after 1945 when ethics began to reconsider areas such as law and, especially, medicine, which saved the field because they returned its relevance and its utility. While I would not quibble with Toulmin's specific historical analysis of the discipline of philosophy, his description of a break between ethical discussion and the practice of life bears no relation to the colossal literatures of ethical ideas and colonial formation in the nineteenth century, nor to the existence of other cultures where the break he observed simply never occurred.

This leads me to my second claim, which is that while the field of the history of ethics has been ahistorical, such narrowness can be addressed through work on the non-West and the pre-modern period. The limitations of the field as it stands can be seen in Reiser, Dick and Curran's massive study *Ethics in Medicine: Historical Perspectives and Contemporary Concerns*, in which the authors wholly ignore cultures outside the West as well as the colonial period.[47]

There do, of course, already exist studies of, for instance, the history of Chinese and other non-Western cultures of medical ethics in both the modern and pre-modern periods, and this book will go on to look at histories of Islamic medical ethics in some detail.[48]

Ethical discussions had always lain at the heart of Arab-Islamic medicine and for Western scholars one of the most remarkable aspects of the history of Islamic medicine is its ethical complexity, and its being structured around ethics of justice and autonomy, as well as those of beneficence and non-maleficence.[49] In the standard narrative of the history of medicine, these 'complementary ethics' of justice and autonomy began to complicate Western medical practice only after 1945 with the rise of 'bioethics' in the United States.[50]

This observation is evidently of critical importance for this book, for when seen from the perspective of the Arab-Islamic world, the ethical encounter which took place in nineteenth-century Algeria was one where one party (the French) operated with peculiarly limited notions of medical ethics, so it was all the more surprising that they purported to represent values of 'civilization' and 'progress' when they were, unknowingly, dealing with a local culture whose medical world depended upon ethical ideas of a richness and complexity not then found in the West. The following chapter will explore the detail and consequences of this situation of incongruity in which a medicalised world was imagined and made in Algeria in such a manner that an existing culture of health which was all around could not be seen by colonists, and local reactions to French medicalisation and its critique could not therefore be foreseen. Across this book my critique of French medical imperialism will both draw on and identify Arab-Islamic conceptions of justice and autonomy in recognition of their true importance to nineteenth-century Algeria.

While it is the case that the history of empire needs to be written into the history of medical ethics, akin to the presentation of themes such as medical experimentation and abortion, we need to acknowledge that ethical questions have been central to the enterprise of the field of the history of colonial medicine. Such works have often begun from theoretical perspectives which presume an ethical duty to attend to the subjectivity of the colonised as much as colonisers – as in David Arnold's writing which draws on the approach of the Subaltern Studies movement – and they have displayed an instinctive willingness to engage in moral critique of the consequences of the establishment of colonial regimes and healthcare systems.

The third weakness of existing literatures in the history of medical ethics is that they have tended to concentrate on the analysis of ethical codes as opposed to moral cultures. It seems clear that this emphasis led to the linear assumption that medical ethics had changed little between the ancient and modern worlds. Cooter identifies an additional difference between historical and ahistorical approaches when he remarks that 'Unlike modern medical ethicists, who aspire to solve ethical problems and to apply their solutions universally, social historians of medical ethics seek (at minimum) to contextualise the construction, colonization, and institutionalization of medical ethical discourses and practices.'[51]

Cooter thus indicates that the broader problem of a concentration on codes has surely been that, unlike histories of colonial medicine, histories of medical ethics have made little concerted attempt to understand how ethical codes and practices affected the lives of doctors and patients. The traditional concentration on codes at the expense of cultures is strange when viewed from the historian's perspective, for the notion that the peoples of the past should necessarily codify the moral assumptions which underpinned the practice of their lives is an inherently strange one. It is especially peculiar in a colonial context, where ethical discussions of medicine and its potential political or civilisational efficacy were plentiful, yet such literatures never made any form of reference to formal medical-ethical codes, even when they invoked ethical language as a means of justifying certain ideas or practices.

The fourth and final limitation of histories of medical ethics is that they have tended to concentrate almost exclusively on the work of doctors. This is atypical in terms of the broader history of medicine which, in recent decades, has extended its interest in doctors to a concern with nurses and other health professionals, politicians, soldiers and others whose work had medical impacts, and to the lives of patients. This approach in fact mirrors traditional approaches found in the Arab-Islamic world, with their emphases on patients, as in the work of al Razi, and on the broader social, cultural, geographic and moral environment on questions of health, and I would like to borrow from that tradition in suggesting that this book is a contribution to the history of health and ethics.[52] It is concerned with the lives of the peoples of Algeria in the nineteenth century, their quality of life, the kinds of healthcare they received, its impact upon their bodies, their life expectancies, the manner in which their bodies and selves were described, especially by doctors and in medical texts, and the relationship between medical textual production and people's lives.

In closing this section, let me observe that while the history of medical ethics has been a relatively underdeveloped field, Algeria has arguably played an important part in that story given the role of Algerian examples in Fanon's critique of colonial medicine and what he saw as the ethical flaws of modern Western medicine more generally. Fanon was particularly alive to the ethical nuances of medical imperialism in Algeria, admitting that Western medicine had things to teach Algerians and that it could do good in the country; yet he observed that when an Algerian said 'I believe what you did was good', the colonist heard 'Do not leave us. What would we do without you?'[53] The French, he argued, were always willing to impute more and more significance onto the idea and practice of medicine, such that colonists and Algerians held very different pictures of its purpose and meaning. This mismatch of ideas and of picturing the other formed the basis of Fanon's critique and this book sets out to show both how this particular culture developed in the nineteenth century and the consequences it had for the French and Algerians.

1.8   On writing about ethics historically

How should the historian write ethically? Do historians have ethical duties? How can historians describe the ethical cultures of the past? These are not, I would suggest, questions which have been central to either historical study or historiographical training.

A stress on ethics can, though, be colossally valuable to the historian in the manner in which it allows us to connect practices and ideas. It allows us to conceptualise moments of choice and to follow the language of patients and healthcare workers, for such people are of course strongly drawn to moral language when describing disease, pain, life and death. It can also afford us access to relationships between individuals and structures (why, for instance, budgetary constraints lead to a doctor choosing a particular form of care for a patient) and, it may allow us to conceive a transhistorical approach to the history of medicine, for if a typology of historical forms of ethics were developed we might then have a clearer idea of the connectedness of medical cultures in world history.

Historical ethics can be seen as a means of linking and describing the relationship between ideas and actions, theories and policies and as a means of describing choices, as they are conceptualised in theory and as they are seen to play out in specific lives and acts. It concerns itself both with intentions and outcomes, and with the relationship between aims and results. I would also suggest that it helps the historians in one of their primary tasks which is the identification of the strangeness of the past (as well, perhaps, as the strangeness of the present).

Critically, it also allows for nuance in historical fields which are often most deserving of complex judgements. An ethical approach, for instance, to colonial medicine allows us to describe more carefully that mixture of 'good' intentions and complex consequences, which we find in Algeria and many other cases, where it is clear that lives were saved and diseases eliminated in the colonial state, whilst new health risks and epidemiological consequences were introduced in a systemic fashion. It can also help, as we have seen, to move beyond the notion that the field of medical ethics is uniquely concerned with the actions of doctors.

While it is the case that historical thinking about ethics has been remarkably under-theorised as compared with cognate disciplines such as anthropology, it is clear that the so-called 'ethical turn' has had an impact on the discipline, most specifically in fields such as the history of empire. Subjects such as the history of war and atrocity in the twentieth century have benefited from ethical approaches in work by Burleigh and Glover, while in colonial history writers such as Young, Davis and Todorov have written works which begin from ethical premises.[54]

This is not a book which depends heavily on existing theoretical approaches but the influence of the work of Davis and Todorov should be clear. Davis's contention that a combination of an ethical drive and the linking together of previously discrete branches of colonial history – such as demography and climatology – lead us towards what he calls 'the secret history of the nineteenth century', has convinced me of both the necessity and possibility of reconstructing the moral world of health in nineteenth-century Algeria and the impact of that world upon the populations of the country. His more general thesis – shared also in recent work on Indian demography and the history of colonial medicine – that historians have tended to underestimate the malign qualities and lethality of nineteenth-century imperialism also fits with the picture of colonial Algeria which I seek to present in this book. We ought to note, however, that while the 'secrets' of this history may have been occluded from historians' views, they were secret neither to critics of empire in the period of decolonisation, such as Fanon, nor, as we shall see, to Algerian writers of the nineteenth century.

The importance of the work of Todorov has been twofold: first, he developed the field of moral history from an unlikely place, in that it was not immediately apparent why a literary theorist working in structural linguistics should begin to concern himself with the ways in which subjugated peoples were alterised in the texts of Columbus and the European explorers.[55] Second, the range of Todorov' subsequent writings has stretched from the early modern *Discovery of America* to the atrocities of the twentieth century, and in developing such work Todorov has evidently begun a comparative study of historical ethics, focused on the human encounter, across the modern world. He helps us to understand the importance of ethical cultures and the need for their reconstruction if we are to understand our past.

Thinking about Algeria, we might ask to what extent did and could doctors operate outside the major lines of the moral enterprise of French colonialism? In a study of contemporary French medical ethics, Malherbe makes an important distinction between Roman-French law and culture – which privileges codes, the centre and philosophy – and the Anglo-Saxon common law tradition which privileges locality, law and individual precedents, suggesting that such traditions played an important role in structuring French thinking about medical ethics.[56] The relevance of this view to my broader argument is made apparent in Weil's critical note on law and ethics in Algeria, for he suggests that 'never before had a regime introduced such a great level of confusion between the words of the law and the lived reality of lives, where terms such as "nationality" and "equality" had been emptied of their meaning'.[57]

This was the ethical bind into which the French placed themselves. In medicine, as much as in law, they were desperate to communicate the idea of ethically driven, commonly applied systems which would always override local circumstances, when the true outcomes of their actions were so often diametrically opposed to such uniformity as ideas, principles and commonalty were sacrificed in the name of expediency, confusion, power struggles and theoretical models which superseded the original ideas in play (such as racial categorisations for instance, which invariably supplanted universalistic expressions of ethics in both medicine and law). In this sense, principles founded on ethics were fantasies, acts of theatre which failed to understand that the short-term applause they received from an audience of their peers, masked the long-run consequences of the responses of other readers of their play, such as Hamdan Khodja and Abdel Kader ben Zahra.[58]

That the French may not have set out to harm Algerians in the medical encounter of the nineteenth century is in some senses immaterial when set aside the medical realities induced through the spread of new diseases, the destruction of traditional ecologies of production and health, the racialising of healthcare, the imposition of new tax burdens, the confiscation of land, the insertion of Algerians into a capitalist economy and the persecution of local healers. One of the most striking aspects of nineteenth-century European writing on Algeria, in public and private forms, is its complete failure to attend to the consequences of imperial policy-making, as though intentions were all that mattered in the world. I hope this book's use of ethical discussion as means to considering both intentions and outcomes acts as a corrective to such writing, for the nineteenth century lie of there being forms of ethics without consequences needs puncturing.

The next chapter of this book looks at the creation of an idea of colonial medicine in Algeria and the ethical ideas which underpinned it, while Chapter 3 further explores the origins of moral thinking about health in the colony through a consideration of humanitarianism. That chapter also begins the work of critiquing the practical and ethical consequences of the beneficent French idea of colonial medicine, which continues the second chapter's study of the ways in which the medicalising idea could and could not be implemented in the colony. Chapter 4 moves on to confront the question as to why discussions of extermination were so prevalent in medical and political literatures on and from Algeria, analysing a massacre at Dahra in this context. Chapter 5 then builds on the work of Ageron, Lorcin and Rey-Goldzeiguer to assess how we might talk ethically about the connections between war, revolt, famine, disease and health in the period 1868–72.

The final chapters of the book look at both how ideas of medicine and the realities of healthcare were described by Algerian-born doctors working in colonial medical services in the last decades of the nineteenth century, before concluding that we ought to see the medical encounter in Algeria as a tragedy structured by the poverty of French medical-ethical thought in a country with a rich tradition of complex ethical understandings of health.

2
On the Idea of Medical Imperialism

2.1 The pygmy's critique

He is a pygmy who knows not whom he dares attack.[1] His accusations as to the murderous qualities of Clauzel and the French army are impudent in the manner in which they ignore the horrific manner in which his countrymen massacred our troops at Blida.[2]

> It is true that the French soldiers had been irritated by the local population, and wanting to avenge the deaths of their horribly-mutilated comrades, had committed a few excesses. These were simply the unfortunate and inevitable consequences of war. They deserve to be deplored, yet we must admit that they were provoked by the inhabitants of Blida, who broke their word and who were amongst the first to commit atrocities against French women and children when presented with the opportunity.[3]

So, in 1834, wrote the anonymous author of the Refutation of the Work of Sidy Hamdan ben Othman Khoja entitled *A Historical and Statistical Glimpse on the Regency of Algiers*. Khodja's *Historical and Statistical Glimpse* had appeared in 1833 and was better know by its Arabic title, *The Mirror*. It was that title, I suspect, which helped to crystallise the anger of his anonymous respondent, for mirrors, of course, can propose an honest view of things which not all might like to see; mirrors can act as an objective third party in disputes between groups who see the world differently. Khodja had dared to question the moral character of the French occupation of Algiers and while Chapter 4 will look more specifically at the massacre and the razzia as events in the history of health and ethics, here I wish to concentrate on his notion that there existed great differences between the colonist's view from his mirror and the reality of the events it depicted.

Khodja was described as being a pygmy because he lacked the sense of perspective and foregrounding which the French mirror on Algeria provided. He could not or would not appreciate the global, historical context in which events such as the massacre at Blida ought to be interpreted. This lack of vision was ultimately an ethical failing for if Khodja had only been able to see things as his anonymous critic had seen them, then he would have realised that the 'few excesses' committed by French troops were framed by a civilised ethics of war (as well as a redemptive narrative of history), whilst the barbarous acts to which they were a response were clear proof of the innate savagery of the peoples of Algeria. In acting as they had, the civilians of Blida had revealed that the Maghreb was a distinct moral realm in which divisions between combatants and civilians could not truly be said to exist, which was of course an idea which would endure for the duration of the life of the colony.

If moral claims can be mendacious then it is clear that Khodja's repudiator could be described as a liar. The blunt fact was that the French women and children massacred at Blida almost certainly never existed in 1833 (it seems more likely that they are a projection of French guilt at their massacre of Algerian innocents), while the author's abnegation of human equivalence between Frenchmen and Algerians is a telling expression of a central conceit of the means by which Frenchmen came to view Algeria. Algerian resistance to the French occupation of their country was quite clearly judged by a set of standards wholly different to those which would have applied had the case in question been the defence of France against a foreign aggressor.

Khodja was a crucial witness for the manner in which he rapidly understood the way in which French imperialism came to be framed morally and how critical such ethical notions were in terms of the creation of the idea of a beneficent, civilisational conquest. It is to our great advantage therefore that we possess not only Khodja's text but also a contemporary colonial response which neatly identifies those parts of *Le Miroir*'s arguments which were seen to strike most closely at the heart of France's imperial idea.

As is the case for much of the history of nineteenth-century Algeria, and empire more generally at that time, we possess few Algerian sources, so the existence of a tract which sees its duty as describing the manner in which French imperialism was becoming structured in Algeria is of great importance. Khodja was a powerful man who had been a rich merchant with close ties to the Ottoman regents of Algiers before 1830, and in the early days after the conquest he had developed good relations with the French, for whom he was a useful contact given his mastery of Arabic, Turkish, French and English, as well as his insights into local politics and power.[4] *Le Miroir* was in fact written in Paris, to where Khodja had travelled to make contact with Ottomans and French anti-colonialists, though I see no reason to view his work in the manner of St John, the British consul at Algiers, or the historian Charles-André Julien, who saw him as something of a mouthpiece for French liberals who wished to quickly end the Algerian experiment.[5] Julien notes that Khodja's argument appeared to depend more on references to Constant, Grotius, Tacitus and the Western liberal tradition, than it did to

'Islamic thought', though this rather tendentious claim ignores the fact that the well-educated Khodja was wholly at home in both European and Islamic cultural *milieux*, and that it evidently made rhetorical sense for him to impugn French motives on the bases of their claims to founding a liberal empire (leaving aside those ideas of an ethics of justice which were so well developed within the Islamic tradition).[6]

That stress on justice was apparent in his anger at the 'bad faith' of the French who had reneged on the treaty agreements with Algerians which Khodja had helped to negotiate, and for which he now felt responsible as he saw his countrymen being denied the rights and dignity enshrined in such pacts.[7] While it was true that Khodja lost much of his own wealth to the French in a very short space of time, his resentment was always expressed on behalf of the Algerian people rather than himself, and centred on his claim that the liberal idea of French empire concealed its despotic, brutal and thieving nature in practice: 'all these horrors which are committed in the name of France and her representatives in Algiers, lie contrary to the principles of liberalism and, indeed, true civilisation'.[8]

The violence of the anonymous refutation of Khodja's work can therefore be explained by the manner in which Khodja had skewered the moral, civilisational lie on which the empire was founded. In 1834, it was essential for France to win this battle in the realm of ideas. If France was able to establish this notion in an epistemological sense as the very basis for the existence of the colony, in later times such an idea could simply echo as a memory of a foundational imperial truth: that the colony was established on the basis of goodness and must therefore remain a good thing. Khodja's injunction that the moral truth of the creation of Algeria be remembered for all time was of great danger to imperial ideologues, most especially because it borrowed their own language as a means of critiquing their actions, whilst also offering a reminder that the people whom France treated so brutally also possessed a 'civilisational' heritage. Across this book, we will see that Khodja was broadly successful in his goal in ensuring that the morally beneficent civilisational claim to Algeria was regarded in both French and Algerian sources as inherently unstable. In this chapter, we will see how ideas of medicine were used to shore up its validity, and in Chapters 6 and 7 we will look at the ways in which the careers of Algerian-born doctors tended to show a movement away from trust in French beneficence to a resistance to French rule based on an understanding of the kind of moral duplicity and consequences identified by Khodja. Contrary to Julien's claim, Khodja's rejection of the French idea of Algeria, and her moral claim to that idea, was founded on the notion that there was a higher religious ethical realm, for Khodja contended that 'God would scorn those who sponsored the suffering of others'.[9]

Khodja's list of proofs in support of his central assertion was a long one, in which questions of medicine, health and welfare played a leading role. He alleged

> that France reneged on her promises of a fair treaty-process...that the goods and property of locals have been pilfered by the French...that the French army has behaved barbarously and despoliated the land...that it was the victors who were savages and the defeated people the exemplars of civilization...and that the charitable donations and legacies held by religious foundations had been seized, with a consequent collapse in the welfare base of the city and a massive increase in misery and poverty.[10]

Khodja's last accusation was in fact supported by French sources which celebrated the fact that the cost of the invasion had been subsidised by confiscations, most especially from charitable and religious foundations, for such acts offered a powerful repudiation of the argument that the conquest of Algeria would be a costly enterprise which would offer France few economic gains. Yet the real significance of such looting was that it revealed the manner in which traditional structures of health and welfare were carelessly dismantled, serving as an exemplar for all such cases outside the capital about which we know little, and revealing of the cynical forging of an idea of a morally beneficent imperial medicine which would care for an indigent class whose impoverishment had its origins in the policies of the empire.

Indeed, while a proportion of charitable funds were re-released by the French for their original purpose, it was typical of the rhetorical model of medical conquest, that moralistic reasons centred on health should have been deployed as a justification for such confiscation:

> The revenues raised for Mecca and Medina are too great for the administration to be able to leave them in the hands of our enemies the Moors, who might use them to nourish and foment trouble. A part of these monies was used to support pilgrimages to Mecca and these are journeys which we have the greatest of interest in discouraging, since each wave of pilgrims brings with it plagues and disease on its return.[11]

The central claim of Khodja's *Miroir* was that on arriving in Algiers, France set about constructing a fantasy as to why she was in Algeria and the kind of state she was creating. This moralistic idea in fact came to take precedence over facts, for higher orders of reality were appealed to in the ethically beneficent *mission civilisatrice*, yet the reception of the idea's truth was hardly as universal as its advocates might have hoped, in either France or Algeria.

This chapter begins with a consideration of the *milieu* of health in Algeria which the French confronted on their arrival, before moving on to look at the construction of the idea of medical imperialism in much more detail, and then the practice to which the idea of medicine was put in the first four decades of the colony, supplementing existing literatures with the consideration of new sources from the nineteenth century. It is important that we keep the words of Khodja in mind, for I shall rely on some of his basic claims as a means of describing the idea of medical imperialism, most especially where we are bereft of Algerians's commentaries on the manner in which their conquerors sought to medicalise their society.

2.2 The picture of health in Algeria

Our early sources, which are almost uniformly Ottoman or French, would seem to suggest that the medical world of Algiers in 1830 was much impoverished as compared with its glory days in the early medieval 'golden age of Islamic science'. As Villot remarked, with specific regard to medicine, 'How far we are from the days of the Arabs of Cordoba!'[12] It was, of course, in the strategic interests of both colonial powers to stress such a narrative of decline – in which a redemptive role could be played by an imperial occupier – and one might indeed argue that the French project of medical imperialism was in some ways modelled on its Ottoman antecedent. Both sought to express ideas of medicine as moral justifications for the healing of a sick culture. We will go on to look more closely at this *mélange* of medicine, history and morality which underpinned the political and military realities of invasion and occupation.

Saidouni's recent study of health in Ottoman Algeria, which is based primarily on French and other European sources, presents a relentlessly negative picture of the health of the Algerian colony at the end of the eighteenth and beginning of the nineteenth centuries. This is somewhat frustrating, for although the unanimity of his sources seems convincing, it is evident that these texts were predisposed to describe an uncontextualised picture of misery and degradation which included no comparative reference to the health of European societies or the complexities and details of local practices of healthcare in Algeria (or, rather, that thin coastal littoral which constituted 'Algeria' at this moment). A more interesting aspect of Saidouni's work is the fact that he connects the political environment of Algeria, and relations between subjects and their masters, with the poor management of public health by the somewhat disconnected Ottomans.

This was especially evident in the instances of plague which struck Algiers in this period, with 16,000 dying in the epidemic of 1784–88, 12,000 in the period 1793–1804 and 13,330 fatalities between 21 June and September 1818.[13] The plague diary of L.P. Vicherat estimated that the population of Algiers, which had stood at 85,000 in 1785, fell to around 50,000 by the end of the eighteenth century.[14] Poor management of the movement of trade in and out of Algiers, including the medical management of pilgrimages, and a lack of public health planning and quarantines, undoubtedly deepened the impact of these plagues, though we ought to note that plagues remained common throughout the nineteenth century and that, in spite of the supposedly coherent structures of the French medical state, Algiers was still devastated by plagues in the 1920s.

Saidouni also asserts that 'in terms of medical knowledge, it is probable that most of the population were unaware of basic principles of hygiene. Rats and parasites were not viewed as vectors of disease transmission, and people relied on rudimentary cures such as garlic and onions.'[15] Nevertheless, Rehbinder, a German visitor to Algiers at the end of the eighteenth century, did comment upon 'the pharmacy of Algiers which contained an impressive array of pots, glasses and containers, full of a heterogeneous collection of ingredients'.[16] Thinking ahead to the large numbers of French ethnographic works of the nineteenth century which uncovered similar evidence of massive pharmacopoeia, along with knowledge of complex surgical techniques, such as trepanation, amongst remote tribal groups in Algeria, one is tempted to disallow much of Saidouni's argument.

While it certainly seems true that the health of Algiers and the major coastal cities was consistently poor on account of the Ottomans' inefficient management of public health, it does not necessarily follow from this that medical knowledge was universally poor in Algeria. It suited the Ottomans, as it would suit the French, to infantilise the Algerian population in this way, but there is as much evidence to suggest that complex ideas of health, which were appropriate to and drew from places where people had lived for centuries, were as common as a narrow reliance on garlic and onions. In fact, sources such as the writing of Hamdan Khodja reveal that while there were parts of what would become Algeria in which certain fevers were endemic, such as the Mitidja, locals had adapted their patterns of settlement and production in such places. Although the Ottomans and French viewed such areas as dangerous and in need of public health works, it is evident that locals were not, as the French thought, climatically adapted to local diseases, but made safe through their own systems of public health.[17] What certainly is clear in reading Saidouni is how very different the situations of health were in the different geographic zones of Algeria – coastal littoral, mountains, plain, desert – and how separate the lives of the peoples of these areas tended to be from one another. It seems quite conceivable, as we shall see, that the French project of making a unified nation state from such disparate regions would have a number of negative consequences for the health of the inhabitants of these areas.

It is unfortunate that we know rather more about the twelfth-century theorisation and practice of 'Islamic medicine' than we do about that subject in the nineteenth century. Since this book looks at the medicalisation of the Algerian colony in the form of sets of ethical conversations between French and Algerian groups, it will be of value to set out some of the principles of so-called 'classical Islamic medicine' before moving onto what we know of its existence in Algeria in 1830.

Five ideas can be regarded as underpinning classical Islamic medicine, the first of which was a stress on geography and 'the importance of ecological conditions to health'.[18] Ullmann notes that this was 'recognised well beyond the medical profession in medieval Muslim society', and we will see that his legacy of the classical period was deeply entrenched in much of nineteenth-century Algeria. Indeed, we might posit that there existed a fundamental divide between the Islamic management of sickness which stressed the relationship between ecology, locality and health, and those European notions which became emergent in the colony where emphasis was placed on the engagement between medical professionals and individual patients.

Second, there was the importance of ethics to Islamic doctors and patients. As Ullmann writes:

> Medieval Galenism retained the strong tradition of medical ethics, perhaps Galen's most enduring legacy to modern medicine. Thus idealism was conveyed, particularly in the *Hippocratic Oath*, as well as in the *Nomos* and the spurious *Testament of Hippocrates*. Many Islamic authors recapitulated and discussed the *Oath*. The Hippocratic ethic is clearly reflected, for example, in Ibn Ridwan's autobiography, where the seven qualities attributed to a doctor are a paraphrase of the *Oath*, and in his other works. He also wrote commentaries on the *Nomos* and the *Testament*. Moreover, the deontological works of the Islamic doctors emphasised the free treatment of poor patients. A concrete expression of professional charity to the poor was the hospital.[19]

The hospital, the third structuring feature, was of course an invention of Islamic medicine, and like much else in classical Islamic medicine, such as the revival of Galenic and Hippocratic traditions, it formed a part of chain of the transmission of scientific ideas, which began in Greece, and which saw the movement of Islamicised neo-classical works move to Europe in the late medieval period.

With regard to my fourth feature – religion – we might note that most nineteenth-century French observers believed that the dominant feature of the Islamic view of health was its idea that disease was a punishment from God. Sickness therefore needed to be borne stoically, for the acceptance of punishment was of greater importance to Muslims than any search for cures. This reputed fatalistic attitude towards health was endlessly remarked upon, yet it was in fact completely unrepresentative of Islamic teachings on sickness and health. It was, after all, Prophet Mohammed who had encouraged the development of science in Islamic societies, which had been a radical proposition to make in seventh-century Arabia. Historically, many subsequent Islamic rulers saw themselves as having a duty to promote experimental science and medicine, for many of their subjects would view the quality of their rule through their promotion of science and culture as expressions of religious faith. As Ullmann notes, 'In comparison with medieval European medicine', this scientific approach and 'the promotion of Galenism helped to establish a nonmoralizing and non-condemnatory interpretation of diseases and their victims in Islamic society, such as lepers',[20] for 'as an intellectual tradition ... Galenism sustained a rational and secular approach to the fundamental questions of health and illness'.[21] In other words, the French picture of fatalism as an expression of irrational belief in Algeria was a complete misrepresentation of one of the dominant views on sickness in Arab-Islamic culture. The determination of French colonists to establish a rigid opposition between secular European medicine and faith-based Islamic healing was based on a fundamental misapprehension of the existence of both secular, scientific and faith-based, popular forms of medicine in the Arab-Islamic canon, and of the traditional dominance and status of secular approaches in much of the Arab world.

It is true that within some branches of Islam there were major objections to rationalism, as evinced in cults of saints, superstition and, in the case of al-Ghazali and some branches of Sufism, a rejection of the rationalistic scientific and medical tradition.[22] It is also the case that the strong, centralised regimes of the medieval caliphs, which acted as vigorous promoters of science, had waned by the modern period, and that localism probably tended to encourage superstition over rationalism, but the fact is that our sources were so predisposed to only describe superstition and to ignore rationalism that we do not know what blend of these approaches existed in Algeria. Clancy-Smith's work on the veneration and funding of secular science in a religious community in the Algerian Sahara at the end of the nineteenth century suggests that we need to acquire a much more detailed knowledge of this subject across the modern period.[23]

Fifth, we may note that within Islamic medicine, as was the case in philosophy and so many other areas of life, there was an instinctive openness to the traditions of other faiths and places which Islam encountered as it moved westwards and eastwards. In medical terms, this expressed itself both in the adaptation of ideas from Greek, Roman, Hindu and Semitic traditions and also in the status held by, for example, Jewish, Christian and Zoroastrian doctors in the medieval Islamic world. Muslim scholars and rulers were well aware of the advantages that could accrue to their societies from the relative lateness of their Prophet's revelation, which goes some way to explaining the instinctive scientific openness and the effective modes of rapid transmission of ideas which existed across the Islamic world. Finally, we ought to mention the division of classical Islamic medicine into a number of well-defined fields, which included pharmacology, anatomy, dietology, surgery and so-called Prophetic medicine. It was this last field – which involved the recitation of verses of the Qur'an, the use of holy amulets and so on – which was invariably stressed by French doctors at the expense of other medical spheres.

The question remains as to which aspects of this Islamic medical culture remained in early nineteenth-century Algeria. Saidouni is sure that such traditions had disappeared, writing that 'Arab medicine from the preceding centuries had been almost entirely forgotten, with the exception of works by Mohammed Ben Ahmed Ech-Charif-El-Djazaïri, Abderrezak Ben Hamadouch, Mohammed Abou Ras En-Nasseri, Mohammed ben Rajab El-Djazaïri and Hamdan Khodja.'[24] He also contends that there was a gap between country people's reliance on the *toubibs* or healers and the urban rich who consulted European doctors, which we see echoed in the French doctor Marty's later remark that 'While those who truly deserve the title "doctor" do not exist in Muslim countries, there are a prodigious number of individuals working as healers.'[25] Jean Temporal, a French doctor stationed in Bône in 1841–42, who was a rare Arabic speaker, stressed the great difference between medicine based on the scientific method and medicine based on superstition, and claimed to have had a patient who had revealed to him that 'today the Arabs have a much diminished knowledge of cures as compared with the days when men were better and favoured by God'.[26] This latter claim is of interest for its encapsulation of the three central tropes of the French account of Islamic medical practices in Algeria: that it had declined from its medieval peak, that healing was seen as a judgement from God and that an essential qualitative difference existed between European medicine and Arab healing.

While such views became commonplaces, and served as one of the justifications of the idea of French medical imperialism, we do find some exceptions to the unanimity of their negative portrayal of Algerian medicine. One writer whose work reveals an uncertainty as to whether there was a strict dichotomy between Algerian and French forms of medicine (which in fact replayed a pre-existing debate within France about scientific medicine and empirics' healing, between reason and superstition) was Capitaine Rozet, a soldier and engineer who described himself as 'attached to the Army of Africa as an Engineer-Geographer, Member of the Society of Natural History and of the Geological Society of France'. In the main, his three-volume 1833 work *A Voyage around the Regence of Algiers with a Description of the Land Occupied by the French Army in Africa* offers a negative portrayal of local medicine: Algerians are criticised for their belief that plague was a punishment from God and the lack of quarantining of people and ports,[27] for their faith in the power of divines and the healing properties of the tombs of *marabouts*,[28] and the racial hierarchy of medical provision which existed, with the poor health of the Arabs, Berbers and Jews, and the better condition of the Turks and Moors (a stratification soon to be replicated by the French).[29]

Yet, almost in spite of himself, Rozet was forced to admit that he saw signs of what might be described as an indigenous medical culture. He noted that Algerians were aware of the palliative effects of certain plants,[30] he found a

Moorish pharmacy in Algiers, though he chose not to enter therein,[31] he observed that barbers and 'some other people' would bandage wounds and perform minor surgery (though he claimed they were more likely to kill than to heal),[32] and he found a hospital at the Bab-el-Oued gate (noting their lack of skill in dealing with fractures and their tendency to amputate).[33] So although Rozet was critical of the practices of such medical outlets, which was understandable since his point of comparison was with the very best of Parisian medicine, he does reveal that when he talks elsewhere about there being an absence of a local medical scene, he was really talking about the fact that he met only two resident European doctors in Algiers, and that he was not referring to the varied and vibrant medical culture he chanced on whilst strolling around the town.

There is also something of an irony in Rozet's text for while he castigates Algerians for their contrivance of links between morality and health, he blames the high mortality rates suffered by the French army (10 per cent of troops were lost in two months, with 50 men dying each day between July and August 1830)[34] entirely on 'the intemperance of soldiers'. He noted that while the army had been given excellent medical advice before arriving in Algeria (we may question this assertion), soldiers who were brought down by typhus, dysentery and other gastro-intestinal diseases, died because of their drinking. Rozet himself proudly staved off such disease by drinking either a glass of sugared wine a day or four or five strong coffees.[35] He happened to note that officers suffered much less from disease than ordinary soldiers, but he attributed this to their greater moral bearing and their comprehension of the link between good living and disease, as compared with their dissolute men. In making such claims Rozet therefore inadvertently reveals a commonalty between the two medical systems which he suggested were so different, as well as his reliance on a sociological interpretation of health whose stress on class was familiar from the France of his time.

Much further along a scale of acknowledgment of the virtues of Algerian medicine were those few French doctors who saw that their own medical understanding could be deepened through the Algerian encounter. These included Lucien Leclerc, whose 1876 prosopography of Arab medicine was not only a major work in itself but also a potent critique of the poor state of knowledge and lack of skills in Arabic which had dogged earlier works on Arab medicine by figures such as Amoreux, Cuvier, Pouchet, Sédillot, Freind and Sprengel.[36] Of equal interest was J.A. Battandier, professor of medicine in Algiers in the 1880s, who argued that Algerian herbalism should have become better known to the French and could have been exploited by them, for its curative prowess and its economic value. Battandier went on to say that:

> We should not judge the current medicine of the Arabs on the basis of the books of their forebears. Such writers essentially re-edited the Greek medicine of Discorides; admittedly with some important innovations,

though many of the drugs they cited as proof of their erudition were not truly known to them. By contrast, their practical medicine ('thérapeutique populaire') is of colossal scope, with great variety across regions, and is almost unknown to us.[37]

Battandier's sadness that the French had not gained access to this store of knowledge by the 1880s reflected a central reality of the medical situation of Algeria which had been completely unrecognised by French writers of the 1830s and 1840s: namely that a complex local medical culture existed, but that the French could not see it because they did not imagine that it existed, nor did they have the linguistic skills needed to apprehend its existence. It is thus understandable both why the idea of medical imperialism came to be one of the dominant motifs of the colony and why we often find local resistance to French medicine and an understanding of its imperial purpose.

Much more prevalent than Battandier's openness to the possibility that there existed an Algerian medicine, was Grenouilleau's view that in 1830 'Every aspect of medicine awaited its creation in this country.'[38] Grenouilleau saw no evidence of 'a culture of hospitals' and specifically disallowed 'those rare asyla, known as "maristans", which one found attached to mosques, in which elderly who were sick or mad died in horrible conditions'.[39] Grenouilleau displayed an unwillingness to acknowledge the possibility of the existence of other systems of health akin to that of Rozet, revealing the manner in which an idea of medical imperialism which believed itself to be founded on the absence of medical culture was more truly established on incomprehension.

Similarly, varied sets of views regarding Algerians and their history could be found in contemporary literatures, with doctors playing a leading role in developing conceptions of time in the Maghreb which helped to generate a powerful and complex combination of ideas about medicine, morality and history. At one extreme, an extremely positive characterisation of Algerian history was presented by Ledentu in his work of 1845, *The Secret of Administrative Science: To Know Those Whom One Governs*. He told his readers that 'you should not imagine that the inhabitants of this country deserve their bad reputation', allowing that 'while it was true that their habits, morals, arts and sciences' were different to those of the French, his readers should not forget that 'their ancestors were the masters of our forefathers', and that 'Without the Arabs, Europe would have remained in a state of the basest ignorance.'[40]

A representation of 'the Arab' much closer in spirit to that of Grenouilleau above was expressed by his medical colleague Trapani, who was one of the first doctors to publish work on Algeria. In contrast to Ledentu's admonition of attendance to difference, Trapani asserted that 'The Arabs of today have no connection with the ancient history of the country ... the majority of this degenerate population live in complete ignorance of their religion.'[41] While it seemed that the French could not therefore decide whether Algerians were religious fanatics or irreligious decadents, it was the former view which prevailed, in part as a means of explaining the sustained level of hostility towards the French, and in part because Trapani's claim clearly had no evidential basis.

Like many French writers, Trapani relied heavily on metaphors of health and cleanliness in his moral-historical justification of colonialism, alleging that 'the dirtiness of the place was so great that one could barely see the traces of her former greatness', and it was therefore incumbent upon the French to introduce an empire of cleansing and healing to revive the ancient spirits of Algeria.[42] The dirtiness of Algerian society, as read from the text of the city of Algiers, was also evinced in the early French obsession with prostitution in the city. This took the form of the undertaking of detailed censuses of the racial identities of prostitutes, the moral condemnation of their practices and the rapid regulation of their services. Such a set of views was made plain in Dr A. Bertherand's article on prostitution in Algeria in 1856, for while the essay was actually a part of a global survey of prostitution, Bertherand was determined to prove that there were reasons as to why prostitution was particularly inimical to Algerian society. He wrote that:

> The Algerian woman's relaxed morals and orientation towards prostitution were caused by the fact that the Qur'an encouraged the submissiveness of women, the hatred of women in Islam, the fact that they were encouraged to be lazy and servile, the ease of divorce, the promiscuity of slaves, the existence of polygamy, and the hot and exciting Algerian climate. ... The woman of the Orient, more depraved than lascivious, therefore naturally gave up her body for sale.[43]

Dirt, lax morals and confusion needed to be combated somehow, and, like many authors, Trapani seemed convinced that the answers to many of these problems lay in France's revival of the spirits of ancient Algeria, when she had been a part of the Roman empire, and therefore Latin and Christian in sensibility. France, according to Trapani, should 'rejoice that she had been chosen by Providence' to accomplish such ends, for her war in Algeria 'would go down in the annals of the history of the world'.[44] In the invocation of providence here we again see the unthinking invocation of precisely the form of primitivism identified in Algerians, and, like most of the tropes associated with the project of medical imperialism, it was much repeated. Thus, we find another doctor, Eugène Bodichon, declaiming in 1851 that 'Providence has called us to Africa in order to regenerate this country through moralising and unification. Our mandate is therefore to destroy all local laws and customs which stand in the way of this project.'[45]

Such themes also relate to the broader regional context set out in Bourget's work in the history of science on the 'invention of the Mediterranean' in French colonial expeditions to Egypt, Algeria and Morée. Looking at the writings of doctors and anthropologists on those missions, she writes that 'far from being struck by similarities between the two shores of the Mediterranean, doctors who accompanied French armies saw only differences, underlined by their helplessness in the face of fevers and local illnesses which could not be accounted for in their own taxonomies.'[46] Such observations formed one part of a more complex narrative, for just as present difference was stressed, this was contrasted with the idea of a common Mediterranean past, and the potential for colonialism to create a common Mediterranean future. In this manner, France would be able to reconnect that which had become detached from the central body, and it is not surprising that medicine and medical metaphors played such an important place in the story of the sewing together of the body of Mediterranean culture. Nor should we ignore the place which the history of religion evidently had in this story-making, for Christianity was obviously identified as a unifying, civilisational force, whilst Islam was cast as a degenerative influence which had divided Mediterranean society.

Such an argument fed into Pirenne's later thesis on European economic history and its collapse in the post-Roman period, when, in his words, the Mediterranean became a 'Muslim lake', thus breaking down the culture of trade which had been encouraged by the Romans. Like Pirenne's thesis, the nineteenth-century account of Mediterranean historical decline was founded on a profound ignorance of the history of the Islamic world, best seen in its stress on the united character of European thought and the divided nature of southern Mediterranean intellectual history. We might in fact argue that the reverse was just as accurate a description of things in the 1830s. The Islamic world, after all, had remained a stable cultural zone of colossal size for more than a millennium by this point, and while its political shape and forms were subject to rapid change, there were significant intellectual and scientific commonalties in the eastern and western Mediterranean, just as there were between the Levant and India. European thought, on the other hand, and French thought in particular, was riven by a central division between religious and secular modes of thinking, based upon the slow seeping of the consequences of enlightenment thought into European societies. This was evinced in places like Algeria in the opposed views of those who believed that the revival of Mediterranean culture would take place through a Christian mission and those who believed that the apex of European culture which France represented was typified by the secularism of her imperial mission. What was to happen in Algeria was that ultimately no decision was to be taken as to which of these routes was in fact better or more practical, leading to a series of important political and ideological conflicts within colonial society.

It was clearly not the relationship between such narratives and the realities of things which ultimately mattered, for their force did not depend upon a basis in facts. Bourget observes that the construction of the Mediterranean and France's political ambitions in the region were two things which fed into each other,[47] neatly observing in French that 'Dans la définition d'une aire méditerranéenne l'histoire aussi est un enjeu'[48] – 'In defining Mediterraneanness, history was very much at stake.' What is more, the power of this narrative was such that its opposition to facts made no difference to the length of its prevalence on both sides of the Mediterranean.

The genre of 'instant histories' which emerged in the 1840s asserted that the first decades of colonial conquest should be placed in a broader context, which tended to set the vicissitudes of the present not against a backdrop of past centuries but aside a vision of what Algeria was surely becoming at that time. As the anonymous author of the 1845 work *Les Princes en Afrique* wrote, 'for the past fifteen years Algeria has been unique for the interest it has afforded contemporary history'.[49] This history was most definitely a form of literary performance, for she 'offered the most charming and dramatic of spectacles'.[50] Its staging saw a confrontation between two great themes of history: war, with its grandeur of destruction and nobility, and 'civilisation' which was the balm which soothed the horrors of war.[51] It was an acknowledged certainty for this author that 'Algerian civilisation would come to be seen by the world as one of the great achievements of France in the nineteenth century.'[52] This 'beautiful page in her history' was being writ large all around, and what is important to see is how pervasive, convincing and powerful were such notions of the ameliorative power of historical change which was actually contained in the present by a providential nation that had access to a mastery of time and balming civilisational tools.

We are dealing here with a profoundly irrational set of beliefs which described themselves as being the logical outcomes of an enlightenment experience and which together cohered to form the rationale for the subjugation of the Algerian people and the creation of systems of governance which were expressive not of practical goals which related to the needs of the moment, but to the coming into being of an assured future. It is for this reason that I am insistent that we see early French colonial writing on Algeria as a form of articulation of a dream, for it is a conceit of the dream that we gain access to new understandings of time, a mastery of its substance. We now need to track the medical dreams of writers who claimed that they were able to sow the seeds which would generate an inevitable civilisational, historical harvest in Algeria.

2.3   The idea of medical imperialism

Writing in 1849, the French writer and resident of Philippeville, Madame Prus contrasted the realities of life in the colony with the views of metropolitan enthusiasts for Algeria:

> The Military authorities, living on the spot, and acquainted with the real facts of the case, plainly declare 'These gentlemen are visionary enthusiasts, utopian dreamers. On the strength of their vivid imagination, but at the expense of government, they send out thousands of unfortunate adventurers whose sole fate is to perish in misery on this African soil. In vain do they hope to carry out their theory of African colonisation.[53]

The army in Africa were not as immune to visions and dreams as Prus here supposes, yet I find there to be something quite revelatory about the manner in which she identifies the centrality of ideas, rhetoric and theory to the Algerian enterprise. The limitations of visions and utopian thought as the bases of politics were quite clear to the army, but they could not be ignored, for as Prus observes, from moments of 'vivid imagination' came strategies and actions which were to shape the colony.

When we are studying nineteenth-century Algeria, therefore, we are interested in ideas about the colony not simply in and of themselves, but for the manner in which notions which may now seem to us quite fantastical served as the bases of forms of governance, even if, like contemporary critics of such ideas, we can see the great limitations of such fancies serving as a means of forming a new country. Politics in Algeria was partly a case of the development of competing positions based on what Prus called 'the real facts of the case', but quite often it emerged from contests between 'facts' and ideas, or between the visions of rival dreamers.

Here I set out the character of forms of French medical imperialism in Algeria, assessing how such ideas formed a body of thought which had a profound impact on the formulation of policies and on the lives of Algerians. I am centrally concerned with the development of a distinct culture in the 1830s and 1840s, the mores of which would have a deeper significance through Algerian history for they were formative of the colonial state. I tend to think that there has traditionally been an underestimation of the significance of such early fantasies in the manner in which French Algeria was structured. Explaining the configuration of the colony will also involve looking at the way in which imperial forms of medicine had developed in the metropole at this time, and how such things impacted on Algeria, as well as assessing the role which concepts of health played in the formulation of broader ideas of Algeria and what she could be.

Medicine offered a powerful lens through which to view the early colony, not least because Algeria was an extremely dangerous environment for French soldiers and the early settlers. Medicine could facilitate an enterprise which stretched the capacity of the modern state and to some extent the power of medicine became fetishised or even sanctified in the manner in which it served as a means of preserving life. If France was to retain an Algerian colony there also remained the issue of how this could be justified, within French politics, to the French people, to the inhabitants of Algeria and to the wider world. The stock answer to this question, which we read of in virtually every French newspaper, book, pamphlet and letter written on the subject of Algeria in the 1830s and 1840s, was that it was France's duty and destiny to bring 'civilisation et progrès' to Algeria.[54] This moral claim was in part a means of moving on from the rather fantastic origins of the French presence in Algeria – when the French consul had supposedly been gravely insulted by being cuffed around the face by the local Ottoman ruler, Dey Hussein – and the fact that Algeria had been acquired in the last days of the rule of Charles X.

In practice, there were a fairly limited number of ways of France to prove that she was the agent of 'civilisation et progrès', and for very many reasons medicine came to inhabit the role of chief proof of such virtues. While fields such as education, law and artistic culture would be called upon in Algeria as proof of examples of France's beneficent intentions, a problem lay in the length of time which it took to implement them as new systems of life. While it might take years to send an Algerian child through a French education system (in fact only a tiny number of Algerians were educated in French schools in the nineteenth century), hospitals could be established quickly and, in theory, locals could be rapidly healed by doctors possessed of techniques and drugs which were unknown in the Maghreb. This would act as a quick proof both of France's good intentions and the greater sophistication of the culture from which the French came (there was little sense of the manner in which the French and Algerians actually came from medical traditions which were deeply connected, and that modern French allopathic medicine had roots in the Arab world). As Turin remarks, 'The power of healing induced servitude',[55] and her study also shows that the personal zeal and conviction of doctors was seen to be of great value in communicating the use of medicine as a means of conveying the imperial message.

There was a double benefit in such medical propagandising for the fact was that the invading army would need to take doctors with it anyhow, so doctors quickly attained an important role in terms of acting as a point of mediation between the French and Algerians. This was a role which doctors were not to relinquish in Algeria, for we shall see that medical professionals were amongst the most important contributors to debates on the colony in France and in Algeria itself, in most cases taking great pride in their status as the vectors of imperialism; as agents of civilisation and progress.

In the 1830s and 1840s it was not just doctors who were keen to describe a project of medical imperialism, but also politicians, soldiers, administrators, human scientists, writers and artists. Part of the confused nature of the medical systems which were established in Algeria originated from the fact that these different interest groups sometimes had very different ideas of the capabilities of medicine and the best uses to which it should be put. We might also note here that the professional categories listed above were hardly rigid, for it was typical of nineteenth-century French culture that the Algerian colony should have thrown up so many figures whose amateur and professional capabilities fused, so that they were doctor-geographers, bureaucrat-historians-of-medicine or artist-geologist-epidemiologists. These hybrid *savants* were responsible for many of the official reports, scientific commissions, journalism, pamphleteering and books of the 1830s and 1840s which tried to resolve 'the Algerian question'.

Typical of such a group was the doctor Jean-Christian Boudin (1806–67), who was also an anthropologist, Paul Broca (1824–80, a surgeon and anthropologist), Melchior Joseph Eugène Daumas (1803–71, a soldier and geographer) and Baron Jean-Baptiste Georges Marie Bory de Saint-Vincent (1778–1846), who wrote *L'Homme: Essai zoologique sur le genre humain*,[56] announcing that 'the march of science would progress under the aegis of war'.[57] As the anthropologist Quatrefages announced in 1867, 'The scholar walked alongside the soldier, and the alliance had been fruitful, as was well known. In Algiers the example set by the Scientific Commission produced sedulous followers from the ranks of the Army.' The conquest of Algeria was very much a corporate enterprise, therefore, between the army and the nascent human sciences, with, as Lorcin observes, each side (when they were not indeed the same person) approving of the other's researches.[58] In this sense, medicine was not simply a tool of imperialism, deployed by soldiers and politicians, but very much a goal and a driving force in the imperial project, as is very clear from the memoirs, letters and articles of French doctors.

The compulsion to represent Algeria, the excess of description, for domestic audiences, was not of course restricted to medicine and the human sciences. As Said observed the size and scope of the Algerian enterprise was of colossal importance in, 'defining nineteenth-century French consciousness of a vast territorial overseas empire'. Before that, as he puts it, 'France had no India... It had no vast colonial possession in which to roam freely as sole coloniser, and make contact with a long-established alien culture.'[59] The absence of Algerian subjectivity which we find in Orientalist medical literatures is equally apparent in the accounts of Algerian society which were dispatched back home in the nineteenth century, as was any real understanding of either the practise of Islam or the complexity of its history.

Looking at the art of early colonial painters such as Antoine Joinville, Vincent Ziem, Ferdinand Duboc, Etienne Raffort, Jean-Charles Langlois and Etienne Martin, it is remarkable how many of them fix on a sense of perspective located in the middle-ground, close enough to display a fascination with their themes of mosques, tribal warriors, the bay of Algiers, 'les secrets de la casbah', 'luxe, calme et volupté' and the desert, but not so close that they or their viewers might have to personally engage with their subjects. By contrast, twentieth-century French colonial art, as practised by figures such as Pierre Frailong, Armand Assus and Léon Cauvy, some of whom were born in Algeria, displayed a much greater interest in portraiture and other genres of the personal encounter.

Art historians' accounts of such work serve as a useful analogy to the study of medicine. Cherry notes the colossal desire to paint and sketch Algeria even amongst tourists who had previously shown no inclination to paint before going there, citing an English 'handbook on "artistic travel" which claimed that Algeria held "everything that an artist could desire" '.[60] Mitchell's description of the relationship between landscape painting and imperialism might just as easily be applied to medical texts, for he argued:

> That the 'medium' of landscape, though inextricable from imperialism, 'does not usually declare its relation to imperialism in any direct way', representing instead something like the 'dreamwork' of imperialism and disclosing both 'utopian fantasies of the perfected imperial prospect and fractured images of unresolved ambivalence and unsurpassed resistance'.[61]

While it is true that medical texts often declared a much more open affiliation with the imperial project, I believe that they certainly shared the capacity of artworks to contain within themselves both the fantastic, dreamlike character of the idea of Algeria, and an understanding of the tensions which existed in a scheme whereby a dream world was to be made in practice through medicine.

Cherry suggests 'that landscape painting attempted a violent enclosing of land within the frameworks of western aesthetics and visual conventions' and while we might say the same of medicine, I think it is also worth noting that both medicine and art were characterised by an air of innocence.[62] What I mean by this is that whereas the actions of soldiers and politicians, and even educators, were understood to possess a certain moral ambiguity, as seen in the need to justify acts such as conquest and pedagogic imperialism, art and medicine needed no such rationalising in the eyes of either their practitioners or their publics. They were viewed as being, at the very least, morally neutral, but much more likely to be seen to be edifying, in the case of art, or beneficent, in the case of medicine, and as such they are of especial interest to us today for in a sense the ethical gap which lies between their conceptualisation and their practice (or the social consequences which they induced) is greater than that found in the work of soldiers and administrators. What indeed, from a nineteenth-century perspective, could the social consequences of art or the idea of medicine be said to be? Did such things exist?

The narrative of medical imperialism began to be constructed from the inception of the colony. While the pacification of the coastal littoral and the reduction in the horrifically high mortality rates amongst the army were immediate strategic goals, they were invariably connected to grander expressions of the idea of medical imperialism, as we see in the Adolphe Salva's 1832 doctoral thesis for the Faculty of Medicine at Montpellier, in which he discoursed on the role that doctors needed to play in ensuring the possibility of a project of enlightenment through medicine in Algeria:

Who is the man who could be indifferent to the fact that this country has a destiny as a home of enlightenment which will shine across a section of the world? Can doctors remain distant, whilst the inherent sicknesses of the place threaten to halt the march of civilisation even more than ignorance and barbarism could prevent its destiny?[63]

It was also the case that at this time the army perceived the benefits which could accrue to themselves if a moral conquest was sanctioned alongside their military endeavours. In May 1836, the minister of war sent a note to General Clauzel, governor of Algeria, arguing that a field hospital for Arabs should be established at Boufarik:

In my opinion the establishment of a field hospital especially for the Arabs would be one of the most productive means by which the French authorities could gain influence over the local population; it would create closer relations between them and us, while it would also spread amongst them the spirit and the benefits of civilisation and slowly dissipate their objections and antipathy.[64]

It is of significance to our broader story that it was the army which so vigorously advocated the establishment of the Boufarik hospital, and which set about defining the practical gains which could come to both France and her military representatives through such ventures. As the general, Comte de la Rüe said in 1836, 'You can do much more for the Arabs with medicine and marabouts than you ever could with cannons and guns.'[65] One hundred seventy-six surgeons arrived with the army of conquest and in the 1830s one of their chief goals was the establishment of civilian hospitals run by military doctors, nine of which were established by 1840; as Lorcin remarks, 'This ability to penetrate into restricted areas [in terms of the lives of locals and through home visits] made physicians valuable scouts in the early decades of occupation.'[66]

The story of the Arab field hospital at Boufarik is also instructive for its quick demise was emblematic of the many tensions which existed in the field of colonial medicine. While, as we have seen, the ministry of war was a strong supporter of the idea of such a hospital when it was established in 1835, it was actually financed through subscriptions from French notables, including both the king and the queen.[67] Yet by 1836 government support for the project was waning in the face of hostility amongst military doctors to this institution which was seen as a challenge to their local hegemony. The hospital closed in 1836 and was reopened in 1837 under military control; while it had succeeded in implanting the idea of civilising indigenous hospitals, it had also revealed the desire of the military to maintain a monopoly on both healthcare and relations with locals.[68]

Many of these ideas can be found in a tract written by the Maréchal-de-camp B. Létang in 1840 on *The Means by Which French Domination Will be Assured in Algeria*, which I think demonstrates the manner in which a whole cluster of tropes, metaphors, assertions and intimations could coalesce into an idea of medical imperialism which enjoyed almost unanimous support amongst imperial interest groups in French Algeria. This is worth mentioning because medical imperialism was rare in this respect since, as I have suggested, most aspects of French policy in Algeria were subject to deep contestation in France and within the colony.

Létang saw colonial doctors as 'serving both humanity with their science and also our own interests, for the Arabs would accord them the respect which they were predisposed to give to professional healers'.[69] 'While the influence of these new missionaries was initially to be restricted to local tribes', according to Létang, 'soon their mission would be extended and once the ignorant Arabs heard that some of their numbers had been cured, they would conceive of this as a miracle, and flock to our doctors.'[70] This is an especially blunt account of one of the other central conceits of French medical imperialism, which was its certainty that the lack of medical knowledge amongst locals was total and that medical cures were always associated with religion and the miraculous, thus making especial sense for France to conceive of doctors as missionaries. As we have seen, such ideas were mistaken in both their assumption of a lack of a local medical culture and their insistence that medicine was bound to religion in the Arab-Islamic tradition. Where medical imperialism was construed as a conversation with local culture, it was therefore stymied by the falseness of its picture of its interlocutor.

The failure in these early decades of the French to answer what would come to be seen as 'the Algerian problem' – what was Algeria for? How should she be related to France? How could France civilise Algerians? – was revealed in the frustration displayed by writers in the later nineteenth century who made the point that a lack of success in resolving such problems had left France in a situation from which it now found it difficult to extricate itself (particularly, as we shall see, since some interest groups, such as the army and colons, had established deep roots in Algeria, which it might be argued greatly benefited from the failure to settle such basic problems).

As Jacques Kob wrote in his 1880 book *Algeria: A Practical Solution towards a Step Forward*, 'The aim of these lines is to offer a solution to the Algerian problem, which is in danger of lasting another forty years, because of the fact that we have allowed an excess of theorizing to cloud the essentially practical nature of the work in hand.'[71]

Such sentiments are important to my argument because they show that in the nineteenth century there existed a debate within French writing on Algeria regarding the 'excess of theorizing' about the colony and the negative impact that this had on the lives of those who lived there. That effect was evidently also felt by French taxpayers who footed part of the bill for a place which it had been believed would increase the wealth of the metropole. Kob's ideas are also reflected in the title of Lieutenant-général Le Pays de Bourjolly's 1846 work *Considerations on Algeria, or Facts Opposed to Theories*, in which he suggests that:

> Of all the questions of the day, that of Algeria remains both the most serious and the most interesting. It is of increasing importance for the fact that it affects our domestic politics, our foreign policy, the prestige of our army and our budget. Our possession of this African territory affects our national honour, and the lives and wealth of our subjects. Since the conquest of Algeria, the two houses of parliament, newspaper columns, magazines, pamphlets and books have not been able to keep quiet on this subject. Self-appointed specialists, publicists and historians have taken up their pens in order to shed some light on the many faces of Algeria; aggrandizing the question from its tiny beginnings to its now great status.[72]

A cultural industry had therefore built up to narrativise Algeria even before it had become a fixed and stable French territory. Le Pays de Bourjolly suggests that this fantasy of Algeria and what it could be was somewhat out of control in the 1830s and 1840s. We shall see later in this book that while the experience of fighting for more than four decades to secure control of Algeria led to some disillusionment and realism within the metropole, important aspects of the early fantasy of Algeria remained right through the colonial period (and indeed some remain today in both France and post-independence Algeria).

The degree to which there was a dispute between fantasists and realists is seen in an exchange of letters in 1835 between the politicians Passy and Renault, in which Passy was opposed to the maintenance of the colony, whilst his interlocutor, Renault, interestingly tried to cast the pragmatic, budget-conscious Passy as a fantasist:

> You sir are the most redoubtable opponent of the Algerian colony, your antipathy towards Africa has been revealed on many occasions, and your post as the budget spokesman for the minister of war each year allows

you to prick the fantastic bubble ['de rompre une lance contre ce fantôme']
of which your imagination is terrified.[73]

Renault suggests that if only Passy had been able to allow his imagination
to carry him away as had the other advocates of Algeria, then he would have
been able to understand that its conquest was a romantic, poetic, moral
pursuit that could not simply be judged or evaluated on pragmatic, logical,
financial terms.

What, therefore, I think we see in early writing on Algeria is the commonalty of a set of ideas about medicine, morals and history. While writing
about medicine and health evidently had some connection to immediate
issues pertaining to the protection of settlers and soldiers, such dangers cannot explain the great stress placed on questions of health in the generation
of a very specific cultural expression of the rhetorical, and soon to be actualised, power of an idea of medical imperialism.

We will go on to see that this system of morally beneficent welfare was
deeply flawed in its application for a whole series of reasons – poor planning, its racial basis, a lack of appetite for practical application, competing
providers in medical markets and the like – but it is quite apparent that the
importance of the idea of medical imperialism extended well beyond its
notional applicability. This idea served as a form of proof that the conquest
of Algeria was a beneficent act in support of global civilisational ideals, in
which the Mediterranean would be healed and history reordered after
millennia of disruption.

The special role which medicine played in Algeria as compared with
France's more general historical aspirations to act to revive Mediterranean
culture are revealed in Lepetit's comparative study of French scientific
expeditions, under the aegis of the colonial military, across the nineteenth century. He notes that the scientific expeditionary force in Algeria
had much less independence than had been the case in Egypt or Greece
and that its leader, Bory de Saint-Vincent – whose slogan for the service
was that it promoted 'the advancement of science under the auspices of
war', found himself constrained by the restrictions on his team's movement imposed by Bugeaud.[74] In addition, the 'scorched earth' tactics of
Bugeaud greatly diminished the possibility of successful scientific investigation, in terms of what was destroyed and the hostility engendered
amongst local populations. Bory de Saint-Vincent argued that while the
expeditions in Egypt and Greece had been explained to locals as a part of
the French desire to spread learning, and as aspects of wars of national
liberation, in Algeria no such claims could be made, and instead the only
hope of inserting his team into the countryside was through medicine,
though he disparagingly noted that this led to 'the global project of
Enlightenment being boiled down to the application of a set of skilled
techniques'.[75]

Medicalisation was a dream and was itself a part of a broader civilisational, historical fantasy in which a brutal occupation would come to be seen more generally as a good and moral thing. In this sense, the idea of medical imperialism in the early decades after the conquest was a form of moral salve for forms of guilt which the occupation of Algeria generated. We will see why that guilt was so great in later chapters as we review themes such as massacres and mortality rates amongst Algerians, but it is of critical importance that we appreciate that the narrativization of Algerian society was a powerful form of world creation. A colony is a place where the world can be made anew and the process of actualising such renovation is bound to have profound effects on the lives of both settlers and indigenous peoples.

2.4 The administration of sickness

Lapeyssonie's rather upbeat recent history of French colonial medicine had surprisingly little to say about Algeria's place in this story, but his work is of interest to us for the manner in which it lays out a description of the administrative structures which he believes were preconditions for successful colonial medical systems. He identifies what he calls the 'three pillars of a wise organisational strategy': 'agreement in the conception of a plan, a united spirit in its implementation, and relentless perseverance.'[76] According to Lapeyssonie, 'It is only in well-structured organisations, especially those in which participants have had a shared training, that these three principles can be put into place.'

It is perhaps unsurprising that Algeria features as a lacuna in Lapeyssonie's work, for one might argue that in the nineteenth century we see none of his preconditions for medical colonial success in the colony. We have already seen that although the idea of medicine united a disparate set of forces in France and Algeria, there were major disagreements about what this should mean in practice, as there were with all other important questions about the future of the colony and the manner in which it should be governed. It does not seem an exaggeration to say that there was in fact no 'plan' in Algeria, or, at least, that the plan changed so often, that it could never be appealed to as a form of constant or organising reality around which other policies and ideas could cohere.

In this section we shall see that right across the nineteenth century, Lapeyssonie's 'united spirit' simply did not exist amongst those who sought to make colonial Algeria, and while divisions and lines of disagreement changed over time, it was always clear that fundamental divergences existed among military, civilian and metropolitan forces. One might contend that each of these groupings was possessed of 'relentless perseverance' but it would seem more accurate to say that, in the sphere of medicine, they obstinately championed their own visions of a medicalised state as a part of larger struggles amongst the colonial classes, rather than displaying unremitting passion for ideas about health as the basis of their motives. It was also quite evident that at the level of administrative and medical training, that colonial groupings did not share a common culture of 'training' for existing divisions and rivalries from the metropole were reflected in the very different means by which professionals were trained to work in the colony.

Lapeyssonie went on to say that 'The good management of administrative services is essential to the treatment of the sick; it is the most powerful auxiliary factor in the success of the practice of medicine, and in itself can serve to save the lives of many',[77] which leads me to a set of questions regarding ethics and the administration of healthcare. Lapeyssonie is surely right that there exists a clear link between standards of care and the general health of a territory and the organisational efficiency of the services which are provided for those populations. What can we say, though, about the consequences for health of medical systems which were primarily characterised by forms of maladministration and competition?

One of the striking features of the idea of medicine, and forms of its practice which we will look at in this chapter, is how rarely patients, and most especially the Algerian sick, were figured in writing about the health of the country. Colossal amounts of time were devoted to the expression of abstract ideas and visions of how systems might theoretically be implemented, but the suffering of individuals or groups is almost completely absent from texts where we might expect such themes to figure largely. Algerians in fact only really appear as characters in literatures on medicine when they reject the beneficent offer of French healthcare, and doctors and administrators then take some time to explain why this was the case.

From the earliest days of the Algerian colony, it was clear that its culture of governance was becoming structured in such a way that it did not fulfil Lapeyssonie's preconditions for the effective administration of health. In 1835 Baron Volland, one of the 'délégué des colons d'Alger' wrote that:

> When we admit that progress has not been as smooth as we would have liked, we need to admit that the administration must take some of the blame. When one thinks that in the short space of five years, military and administrative power has passed through six different sets of hands, should we be surprised that there is an absence of an administrative system, a complete lack of unity of action and vision?[78]

Twelve years later the same points could be made by Citati, an Italian judge who had gained French citizenship in Algeria in 1832, when he wrote that 'The form of administration which has hitherto been applied to Algeria has been provisional, ever-changing and, because of these things, poor and harmful, rather than useful to the country. We speak of "administration" because there is an absence of "government".'[79] Such questioning led Citati to return to the archetypal epistemic question of the 1830s: 'What is Algeria for?'[80] Yet in a sense writers like Citati could be viewed as much part of problem as of a solution, for the pamphleteers, the journalists and the informed citizens, offered up a plethora of administrative solutions for the colony, which in some ways only made the situation more confused. Citati's own preference was for Algeria to become a viceroyalty, whilst others suggested innovations such as a special ministry for Algeria. Le Pays de Bourjolly effectively demolished this latter idea with the observation that the creation of such a ministry 'would have been to simple add more cogs to the already over-burdened administrative machine of Algeria'.[81] The idea of an administrative machine articulated here is a common one in texts of this period and reflective, I think, of the frustration felt in an environment in which it was felt that scientific principles might be applied, yet where administrative practice seemed wholly chaotic.

2.5   The practice of French medicine in Algeria

When one looks at the budgets, laws and statistical data pertaining to the early colony, one of the most striking aspects of French Algeria was how quickly ways were found to finance imperial expansion, most especially through taxation of the indigènes, confiscation of the tithes which they had formerly paid as a form of collective charitable public assistance, through the taking away of their property and through the increases in taxes which accrued to the government through the growth in trade which followed the conquest. While most primary and secondary sources rightly identify the fiscal crises with which the colonial administration had to deal, what they fail to note is that such crises were primarily induced by decisions made in Paris and Algiers to take profits from the colony, rather than investing monies raised there.

Thus, in the first three years of its life, the expenses of the colony exceeded its receipts on only one occasion, and in 1832, there was a massive surplus that was primarily generated through customs revenues.[82] The deputy Renault claimed that one of the essential problems of the colony was the reluctance of the state to spend money and its silence as to the monies it raised in Algeria, which was compounded by its choosing to ignore the huge upsurge in the economy of the Midi owing to its trade with Algeria.[83] Looking at early colonial budgets it is also plain as to why many doctors, and indeed soldiers, felt that there was a significant disparity between the rhetoric of medical imperialism and the sums which the state was willing to devote to such services. Our knowledge of the budget for 1830 is somewhat limited but it is plain that the budget for policing far exceeded that for health, and in 1831 the nine medical budget lines (costs of the 'hospice civil': personnel, construction, drugs, 'loyer du local'; the 'dispensaire': personnel, building, special expenses; 'Etablissement sanitaire' and 'entretien de lazaret') reached a sum of 95,000 francs, whilst police expenditure accounted for 114,000 francs (road- and bridge-building costs stood at 250,270 francs).[84]

Looking at the statistical record of the metropolitan French national budget of 1830 gives us some idea of the place of the military and military medicine in the priorities of government. The expenditure of the state in that year was 353 million francs with a total of 609,807 salaried employees working for the government.[85] About a third of the state's costs and its employees were consumed by the army and the navy. The cost of the army was 92 million francs, which supported 566 'officiers de santé' (costing a million francs), 160 'officiers d'administration', 'adjutants', 'sous-adjutants' of the military hospitals (278,000 francs) and 729 nurses (165,000 francs).[86] It is instructive to compare the average annual salary of officiers de santé of 2000 francs with that of the military police, who made around 2500 francs. In the navy, there were 357 officiers de santé (costing 553,900 francs), making only about 1500 francs a year, and also a combined cost of 140,000 francs for navy hospitals (which included general employees, intendants, gardeners, pharmacists, herbalists, nurses and porters).[87]

As I have suggested, such expenditure was easily covered by the confiscation of Muslim charitable obligations. In his 1883 collection of French legislation in Algeria, Béquet notes that before the arrival of the French such donations were passed directly to religious institutions and legal sanctions prevented the Ottoman rulers, or indeed any rulers, from controlling such gifts, which were destined to be used for the 'easing of the lives of the sick and the miserable.'[88] One of the first pieces of legislation passed by the French in Algeria – on 8 September 1830 – was a declaration that such monies were the property of the state. Béquet observed that while a small proportion of this money did make its way to traditional charitable institutions, most was either used by the colonial state for its public works programme or simply retained for other uses, which was understandably to lead to a long-term decline in such giving, for Algerians understood that such philanthropy had mutated into a form of taxation.

Béquet's collection is especially useful in the manner in which it helps us to understand the ways in which the French adapted such pre-existing administrative structures for their own purposes in Algeria. Another example of such reworking came in the establishment of a formalised hierarchy of indigène administration from 1845, lying beneath the French chain of command. Ideally, the French envisaged five ranks in this system: the *khalifat*, who dealt directly with the French, and who presided over the *aghalick*, the *bachaghalick*, the *kaidat* and the *cheikat*.[89] This system was not applied universally, since there were certain tribes which, for reasons of power or size, could not or would not, recognise the stratifications of this schema, but in a sense the reality of the meaning of such systems is less important than the manner in which the French believed that they were creating local forms of governance which incorporated both French and Algerian social norms. The ranks listed above also corresponded in a rough sense to a set of geographical distinctions, which began with the *douar*, which described a collection of tents, moving on to the *ferka*, the tribe (commanded by a kaïd), a *grand kaïdat* (led by an agha) and a collection of such groups, led by a bachagha or khalifa.[90]

There are two reasons for setting out such distinctions: to understand the context in which colonial medicine was practised, and to see that aspects of French colonial rule did attempt some kind of synthesis with what were perceived as local social norms. Such a move was also apparent in the joint application of Islamic and civil legal codes, but as we have seen in the crucial area of medicine, there was almost no inclination on the part of the French to understand either the practices of local healers or the administrative structures which governed healthcare in the different regions of the country (other than the confiscation of those funds which supported the healthcare needs of the indigent and the helpless).

Let us now move on to consider some aspects of the practice of military and civil medicine in the first decades of the colony, most especially as they relate to the themes of my later chapters. As my introduction noted, those interested in more detailed studies of medical institutions in the colony are best advised to consult the work of Turin, which I do not aim to replicate here.

The specific story of military medicine is absolutely critical to the history of modern Algeria. Over a period of time, military medicine gained a very high status within colonial society, befitting its role as a double protector, through arms and medical care, but in the early years of the occupation, military doctors were somewhat constrained by the pre-existing status of their work within the army. Their standing was not high as a result of the poor organisation of the military medical corps and the perception that medical service lay on a parallel scale which situated beneath the career structure of the regular army. It was not until 1833 that doctors were able to become officers and 'the question of promotion was also much neglected'.[91] Even later in the nineteenth century, Taithe, in his study of the Franco-Prussian War, notes that:

> The low status of military medical men, their fairly dangerous lifestyle and relatively low income did not attract vocations comparable to the enthusiasm for civilian medicine. ... the life of the army medical man, vegetating in some obscure garrison town in France or Algeria, lacked any appeal and did not favour profitable marriages. ... The type of medicine they practised specialised them early into disciplinarian forms of epidemiology concerned mostly with venereal diseases, fatigue, and simulated ailments.[92]

Taithe also argues that the disparagement of military doctors through hierarchical and administrative means was far from an accident, since it reflected the 'fear' which existed within the institution of 'the notorious independence of spirit of army medical officers'.[93] Much of that independence of spirit was in evidence in Algeria, where, for a number of reasons, military doctors tended to break free from some of the constraints which had traditionally been placed upon them, given the critical nature of the military medical function in the Algerian disease environment (and the evident bravery of military doctors whose mortality rates were of course extremely high) and also including the fact that the army led the promotion of the idea of medical imperialism. More important than these two things, though, was the fact that Algeria effectively became a military colony for most of the nineteenth century. In this sense, it was divorced from both the realities of France and the army in France, and able to develop its own culture.

By 1841, Bugeaud controlled around 100,000 troops in Algeria, which amounted to about a third of the French army.[94] The colossal size of that force certainly helps us to understand how dominant the army were in forming the nascent colonial institutions of Algeria, though this was a source of considerable discontent in a series of circles, military and civilian, in both France and Algeria.

A particularly influential critique of the army of Africa was Baron Baude's work of 1841. Baude, who had formerly held the position of commissary of the king in Algeria, argued that the presence of a colossal military presence in Algeria 'imposed huge costs on France and yet offered few advantages'.[95] He compared Bugeaud's army unfavourably with that of Napoleon's in Egypt, arguing that while the latter was characterised by its qualities of being battle-hardened, prestigious and well-organised, the Algerian force was made up of physically weak and inexperienced conscripts, whose health was often deteriorated through sickness and malnutrition.[96] He contended that it would be both more prudent and effective to abandon the massive ground force in favour of a fitter, elite contingent of 20,000 troops, arguing that the indigènes would be much more likely to respect such a force than they would Bugeaud's shabby soldiers and their equally unimpressive horses.[97] In addition to the Napoleonic example, Baude also suggested that France could learn lessons from the history of the Spanish and the Ottomans in Algeria, both of whom had ruled with smaller armies drawn from social elites.[98] As he remarks 'In Algeria we have hardly succeeded in dominating the Arabs with our 50,000 men [he was unaware the true figure was much higher], whilst the locals trembled before five or six thousand Turks.'[99] It must be admitted that there is some hyperbole to Baude's claim here, for the Ottomans had never attempted either the forms of territorial or social extension which were being undertaken by the French beyond the towns of the coast, but his central questioning of the huge troop presence in Algeria was seen to be valid. As we shall go on to see, and as Baude surely knew, the motives behind the size of the French army in Africa were far from wholly connected to the military exigencies placed upon the army, but were bound up with the army's perception of its political and strategic purpose in the colony.

Before moving on to look at military colonialism, let us consider the resourcing of military medicine in the 1830s and 1840s, having already noted that there were great pressures on the service at that time. By 1848, 33 hospitals had been established in Algeria with a total of 13,700 beds, but there was great variability in staffing and complaints by doctors of under-manning.[100] There were too few medical doctors in the service and in 1835, Clauzel was forced to protest against planned reductions in numbers of personnel at a time of a great cholera outbreak which had, for instance, reduced the population of Mascara from 10,000 to 8,500.[101] While there had been 418 military doctors in 1847, this number was reduced to 267 by 1853, which, like many things in Algeria, was almost certainly the result as much of poor planning as it was of budgetary exigencies.

Turin noted that of the 38 hospitals established by 1845, many had begun to treat local patients as well as fulfilling their functions as 'a necessity of war' ('une nécessité de la guerre').[102] Her choice of words here is intriguing in that it is deployed in a context – the spread of military medicine into the civic sphere – which reminds us of Taithe's later reading of military medicine and humanitarianism in the Franco-Prussian War, where an intentional blurring of the lines of both military medicine's responsibility for soldiers and civilians was accompanied by a blurring of the difference between combatants and non-combatants.

E. Bertherand, however, noted that indigènes did not visit military hospitals, unless they were about to die, and all hospitals were seen as culturally suspicious because they were the home of the *Roumis* [Europeans, literally 'Romans'], with 'their differences of custom, religious practices, language, morals, denial of freedom, imposition of discipline and provision of foodstuffs which were viewed as having been suspiciously prepared'.[103] For these reasons, Bertherand contended that when locals did go to such hospitals they tended to try to leave quickly, during their courses of treatment, and did not therefore receive the benefits of French medicine, and that this suspicion of the French military and its institutions in fact diminished their ideas of the potential of French medicine.[104] The absence of any sense of cultural sensitivity or local autonomy is of course apparent to us now.

Bertherand was surely right in saying that in the early decades of settlement, when the potential of the idea of medicine was thought to be at its most potent, it was constrained by the wider policies of the army. Amongst the most interesting of such plans was the idea of Bugeaud and other generals that Algeria would become a military colony, set out in his 1838 work *On the Establishment of Legions of Military Colonists in the French Possessions of*

North Africa. Bugeaud began by arguing that,

> We are in Africa and we wish to remain there, with one group desiring this as a result of their illusions, and another because they recognise that the administration would be incapable of effectively planning either an evacuation of the country or even a retreat to the coast as a sad first stage in the abandonment of Africa.[105]

Bugeaud evidently identified himself in this second, pragmatic camp, evading a potential charge of military opportunism in his assertion that it was not the army which had concocted these fantastic illusions of Africa that had driven the first phase of imperialism in North Africa. The army was in fact simply responding to the realities of the situation on the ground; which was untrue on many levels since the picture of Algeria which Bugeaud was creating was very much as a result of his proactive policies and his broader dream of a military colony, as attested by figures such as Bory de Saint-Vincent.

Bugeaud went on to set out the conditions for a successful establishment of secure colony in Algeria, which included the creation of self-sustaining French provinces and the encouragement of mass emigration from France.[106] Such goals could not, however, be considered without a primary concentration on security, which was to be established by the creation of military colonies at the edges of the empire ('les avant-postes de la colonisation'), where soldiers would farm and settle, whilst securing the borders of the state.[107] Through such a plan:

> Great advantages would accrue to France through the creation of this new people who would open up commercial relations with the Arabs of Algeria and with the interior of Africa. Such a strong form of occupation would be less likely to be challenged by locals and would succeed in a short time. And, by contrast, if attempts were made only at the slow inculcation of civilisation, then we might spend another hundred years in our current miserable and precarious position in the colony, which would also be ruinous for the metropole.[108]

So Bugeaud's vision was founded on just as wild an illusion as that of other colonial theorists. His idea of a new people forming on the borders of Algeria was based on the notion that soldiers possessed special characteristics and that Algeria represented a great opportunity for the army to experiment in making a society in their own image. His threat to his political masters in France was that if he was not granted the freedom to make Algeria in the image of the army, then the army could do little to prevent it from becoming a terrible imposition on the nation for decades to come.

In a sense, Bugeaud's blackmailing of the state was somewhat unnecessary, for by this point in time the size and importance of the army in Algeria was so great that it might have seemed to be more prudent to simply continue towards a *de facto* militarised future for Algeria. Such a prospect seemed even more likely given the continued, and in some cases increasingly coordinated, resistance which the army was facing as it attempted to expand the borders of Algeria. It was indeed alleged by some opponents of the military that the forces of Abdel-Kader were effectively a form of defiance that had been conjured up by Bugeaud and his forces. In the field of medicine, there was plenty of evidence from this time that a sustained level of competition existed between civilian and military providers of healthcare. Bugeaud's intervention, therefore, was intended to try to settle such contests and to formalise the status of Algeria as a colony of the army, which although close to being realised, could not be countenanced by political forces outside the armed services.

In contrast to the scale and organisation of military medicine in the colony, the provision of 'civil medicine', by which I mean both medical care directed at the colons and indigènes, tended to be more fragmented, prone to sudden departures of key personnel, funding crises and liable to be affected by changes in the political and military cadres. It was, as we shall see, also intimately connected to the military medical establishment, which, while necessary, tended to cause major problems in terms of competition between different medical providers, blurred borderlines of responsibility and confusion amongst both indigène and colon patients.

We have heard already of the stop-start history of the field hospital which was established at Bou Farik, and eventually taken over by the military in 1837, which was one of a number of hospitals specialising in indigenous care which were established in the 1830s. For instance, in February 1838, doctors Méardi, Trolliet and Bodichon set up a hospital at Caratines offering free local consultations, while by 1837 civilian hospitals had been established at Algiers, Oran, Bône, Bougie, Douéra, Mostaganem and Guelma.[109] Doctors such as Pouzin and Giscard – 'I offered medical care to many tribes pretty far from our territories, which allowed me safe passage there' – were enthusiastic advocates of indigenous medicine, but there was some suspicion within the army with regard to the idea of establishing a complementary medical service for locals.[110] In part, this came from the prejudices of non-military doctors who did not want to engage with the indigènes, particularly given the fact that it was not remunerative, and also from the resistance of the colons to the idea of sharing medical services with locals (separate consultation times were usually established).[111] Broader divides on policy surrounded the question of who would control civilian medicine and on what scale it would be offered. In 1843, Bugeaud proposed the establishment of free, comprehensive medical service for all Algerians and settlers, but this was rejected on the basis of cost and because of objections from within the army.[112]

Eventually such debates were settled with the formation of the Bureaux Arabes in 1844, which seemed to institutionalise the separation of European and Algerian systems of care, but in practice often meant that smaller and more isolated settlements were often served by one doctor, whose role shifted as he moved in different environments. The regulations governing the Bureaux were signed off by Bugeaud and they make it quite clear that doctors had a duty to serve military officers and their families, locals in the military hospital and the imprisoned.[113] A critical function of the doctor's role related to policing and surveillance, for the rules stated that

> The doctor must assist the officers of the Bureau Arabe when they were fulfilling their duties as judicial police officers. As far as his other duties permit, the doctor should make trips around the local tribes, and should offer the head of the local bureau a daily report on the things which he has seen in the course of his duties.[114]

In other words, Bugeaud and his fellow officers took on board the remarks of early doctors such as Giscard and Pouzin to insist that doctors exploited what seemed to be their unique access to the lives of locals for the benefit of the military occupiers. What Bugeaud had not fully perceived at this time was the extent to which local peoples had understood very well that doctors were used in this fashion, and that a broader rejection of Western medicine and the offer of French culture might ensue from such policies.

In many ways, the work of doctors in the Bureaux Arabes was emblematic of the structure of French administration in the tribal areas beyond the cities of the coast for each of the Bureaux lay at the centre of a so-called 'circle' of territory and tribes, and it was the doctor and his interpreters who were provided with mules and horses to make trips around the territory. It is not hard to understand why locals believed that European medicine was being used to encircle their culture and their way of life. Such trends, though, were more important in the later life of the Bureaux, for it was not really until 1867 that they gained a significant role in more remote areas away from the coast, which had until then remained under the sole jurisdiction of military doctors.

Doctors attached to the Bureaux stressed the great problems which they faced in fulfilling their missions. These included a lack of time and resources, language problems which made it difficult to communicate with locals, the fact that hospitals seemed alien to locals (more likely in fact that there was a disparity between the Islamic idea of the hospital as a care home or a hospice and its European identity as a locale for palliative care and surgery), the difficulties of treating local women, the fact that doctors could do little about sanitation and issues of public health and the immense scepticism of locals as to the efficacy of European medicine.[115] The problem underlying all these things, as Perkins notes, was that 'Most administrators realised that so long as medical care remained erratic and inadequate it would serve little more than a propaganda function by demonstrating "that the authorities are not indifferent to the evils besetting the tribesman".'[116]

In his 1855 work, *Medicine and Hygiene amongst the Arabs*, E. Bertherand also identified the poor organisation of medical provision as the central problem militating against the work of doctors. He found this frustrating since he did feel able to point to a series of success stories in the Bureaux of Oran, Constantine and Algiers, where 45,000 locals had been treated in the period 1847–50. Based on his work there, Bertherand argued that the prescription of drugs such as quinine sulphate, for malaria and other fevers, and opium, for dysentery and diarrhoea, had convinced 'patients of the good care they were offered, inducing in them both gratitude and confidence in France'.[117] Yet there was irritation in such observations for he regarded such care as somewhat isolated in what could and should have been a 'rational health service' that could have built on the great gains of the Bureaux of the cities. He also suggested that such a system could have effected particularly quick changes in Algeria because of the tribal character of its hinterland, and the consequent dispersal of political power, which would have made the domination of local tribes through medicine all the easier, as compared with potential task posed by countries like Egypt and Turkey where there was a centralised Muslim political or intellectual leadership.[118] One might question Bertherand's observation here on two grounds: first, that France was of course very actively trying to make a single national space in Algeria, and second, that the dispersal of power and influence in Algeria could be argued to have made moral conquest all the more difficult than was the case in colonial situations where all-powerful centralised elites could be co-opted by the imperial power.

Bertherand ultimately concluded that the Bureaux would not prove the most effective means of implementing the idea of colonial medicine. The central problem with the Bureau model was that it was under-funded and doctors' chief function was to be a servant to the military rather than to locals and to the ideal of colonial medicine. He cited a Bureau officer named Lapasset who had grown frustrated by the gap that existed between rhetorical valorisation of medical provision offered by the Bureaux, which he believed in with a passion, and the 'skimping on funds which had prevented the full development of the realisation of this idea'.[119] Such a complaint might stand more generally as emblematic of the problems encountered in the colonial medical service from the 1850s, as it became clear how much it would cost, in terms of money, manpower and administrative ingenuity, to implement the idea of medicine which had played such a critical role in the period of the conquest.

Bertherand's proposed solution to this dilemma was the expansion of the number of hospitals exclusively devoted to the care of locals, which he claimed could be cheaply funded through taxation. He noted that such establishments were much more popular with the indigènes than medical facilities which were shared with the French, and quoted from an approving review from the *Revue Orientale* of the 'double importance' of his hospital in Algiers, which served 'both beneficence and political goals'.[120] Arab hospitals, he contended, had the potential to become the hub of the colonial enterprise:

> For these establishments would be placed under the responsibility of local chiefs and from them the work of the doctor would shine across the Cercle. The doctor would make his rounds amongst the sick; he would station himself in the market-place, he would submit to the power of local authorities, and would implement and plan measures of public health, such as the construction of roads, Moorish baths, plantations, water works and so on. He would receive from local leaders news of births, marriages and deaths, and would, in a word, serve as the hub of the civil administration, which he would construct in such a way as to fit with Muslim morals. Only by living and staying with the Arabs, could the doctor get to know the details of their lives, their idea of the mysteries of being, and could thus gain their confidence, convincing those who were credulous and exposing medical impostors.[121]

Yet from what we already know of military medicine in Algeria, we can well understand how such ideas, even when articulated by one of the most influential doctors in the colony, were never to be implemented. What Bertherand describes is a logical, practical extension of the idea of medicine as a form of moral conquest, which had been so enthusiastically supported by the army in earlier times, yet by 1855 his call for an independent health service which engaged directly with locals, outside of the military government, went against the mood of the times (though one imagines that it would have seemed more palatable had it been proposed by a military doctor). In Bertherand's system, the doctor would have been an alternative source of authority to the soldier and his direct, intimate knowledge of locals, and indeed his seeming primary allegiance to them, could not be countenanced by military leaders. The army still needed men like Bertherand and there were aspects of his project which remained appealing, but its scope and the manner in which it actualised the fullness of the idea of medical imperialism were no longer in fashion. Bertherand's increasing isolation from the mainstream of medical thinking in Algeria by this point can, I think, be seen in his advocacy of a translation project of classical Islamic medicine into French as a way of understanding what kinds of ideas informed contemporary opinion, and how these might be altered by European doctors.[122]

In a sense, what one sees in Bertherand's writing is a playing out of a conflict between the centre and the periphery, which should have been familiar to French doctors in Algeria since it mirrored the manner in which a conflict between Paris and the regions structured medicine in the metropole. As LaBerge and Feingold remark:

> Attempts to promote public health reform show the extent of the conflict between Paris and the provinces, the 'princes of medicine' – the elite of Parisian public health – made hygienism the cornerstone of their programme. In the name of national security and social order, they sought to impose their views and practices on the country at large. The provinces resisted. To them such a policy was yet another attempt by power-hungry Parisian physicians to control the provinces by the imposition of centralised order, which, they believed, would intrude into the domain of the family and private life and deprive physicians and families of individual liberties.[123]

The difference, of course, of the Algerian debate was that in one sense Bertherand stood as a representative of Parisian-style medicine, in that he wanted a centralised idea to be dispersed across the territory, but in another sense he represented the resistance of the provinces, for he wanted doctors in the tribal areas to benefit from independence from the central authorities, for they could achieve much more in acting as entrepreneurs of European medicine amongst the Arabs.

In a sense therefore, the medical system in Algeria was moving closer to the complex nineteenth-century French model of competition between different sources of medical authority – branches of medicine, capital and provinces, universities and medical research, governmental and non-governmental medicine – with, in Algeria, the complicating factors of the relationship between the colony and the metropole and the army and civil society. It was between this last pair that an especially great tension emerged in Algeria.

Algeria came to be governed by a so-called *mixte* system, which meshed civil and military power together, but we have seen that in practice this led to military rule up until 1871, and even beyond that point at the edges of the state. There was an extensive campaigning literature from the 1840s and 1850s, largely composed by professionals, such as doctors, living in Algeria, which argued for more civil governance in opposition to the hegemony of the army. Let us take, for example, Delpech de Saint-Guilhem, one of the 'propriétaires délégués de l'Algérie', who, with the support of his fellow délégués, baron de Vialar and comte de Raousset-Boulbon, told the Chamber in Paris that 'In the end all our demands can be summed up in the following formula: give us civil institutions in Algeria.'[124] Even some army officers had become unconvinced that the solution to the future political status of Algeria lay in their hands. Bourjolly offered instructive descriptions of the struggles which took place between different forces in Algeria in his conclusion that 'The conquest by arms has not succeeded, and never will unless we attend to the more important moral and spiritual conquest of the country.'[125] He contended that to allow such a conquest the potential to work in Algeria, the number of troops stationed in the country needed to be reduced, while the number of colonists needed to be rapidly increased.

Looking at the speeches, books and pamphlets of doctors and medical professionals at this time, it is remarkable how unanimous they were in criticising military governance, and how powerful a voice their prestige generated for them in such debates. Warnery, for instance, blamed the army and the military government for failing to implement important public health plans, such as the construction of canals in the Mitidja, which had been mapped out by the engineer Prusse in 1834 and 1836, yet which lay unbuilt when he wrote in 1847.[126] He also contended that the lack of civil hospitals was a result of the army's opposition to civil colonisation, and suggested that if that hostility were lifted, a civil regime in Algeria had the potential to send 120 million francs a year back to France, in place of the 120 million francs that the army cost annually.[127] Writing with Jules Duval, Dr Auguste Warnier argued that 'In their rebellions against our authority, the indigènes are able to distinguish between the colon and the soldier', but the authors contended that the military government saw it as in their interests to obscure this fact, because to admit that indigènes could make such distinctions would be to abandon the *raison d'être* of the 'exceptional' military regime.[128]

2.6   Conclusion

In later chapters, we will see that while colonial competition, and ensuing forms of maladministration and confusion, characterised the governance of Algeria across the nineteenth century, significant changes took place within the stances of those associated with the army and the civil administration. As time went on, the kind of beneficent identification with the needs of the indigènes which we saw in the work of Bertherand began to disappear from the rhetoric of civil authorities, which became more associated with a colon culture which displayed a hatred for Algerians, while the army's prerogatives on the edge of empire arguably led to their having a more entrenched commitment to the idea of medical imperialism than administrators in the cities of the coast. Of course, major divisions existed within these interest groups, and in the next chapters we will move on to look at, respectively, doctors' roles in establishing racist models for healthcare and the army's advocacy of policies of extermination, but it is also important to recognise that shifts took place within this complex world of the formation of a colonial society.

In this chapter, I hope to have shown that, in both secondary literatures and published works from the nineteenth century, the idea of medicine was a cornerstone of the Algerian project, in both practical and ideological ways.

The power of the idea was founded on forms of cultural self-confidence which contrasted the efficacy and world-leading qualities of French medicine with the apparently dismal state of health and medical expertise in Algeria. Historical writing and medical propaganda became allies in demonstrating that medicine could play a vital role not only in protecting the French army in a dangerous environment but also in performing an act of temporal transformation, which was connected to broader French beliefs about the place she was destined to play in the revival of the Mediterranean as a place of prosperity and well-being comparable to its condition in times of antiquity.

It soon, however, became evident that such overarching cultural and historical theses were poor bases for the establishment of systems of medical administration and public health, and this was further hindered by the endemic forms of competition which emerged between, amongst others, civil and military authorities. Doctors played important roles in these debates, but in spite of their frustrations with the gaps which existed between the idea of medicine and the institutional and budgetary constraints which hindered its development, they tended to express a continuing hope in the power of the beneficent idea and a sense of being seduced by the place which their practice and worldview had in the shaping of a new society. The moral certainty of doing good trumped all fears that maleficent consequences may have lain within this beneficent project.

We saw however, that the consequences of this haphazard medicalisation of 'Algerian' society were immediately apparent to Hamdan Khodja. He gave the lie to the untroubled idea of imperial beneficence and it would seem that his work, and the French engagement with his critique of the development of the colony, revealed that France could not have been quite as unselfconscious as to the malign consequences of its civilising mission as it purported to be. Yet of course, there were powerful reasons as to why Khodja's voice was disallowed in such debates. He was after all speaking against the force of history and its progressive impulsion through an idea which would be actualised in an inevitable fashion. That Khodja, unlike his interlocutors, was possessed of a rather unique sense of perspective on these things – for he was familiar with the world of the Ottoman court, the life of Paris and the various strands of Algerian society – could be denied in a second way on the basis of his race and religion, and the inevitable skewing of his priorities which ensued from his inherited fatalism. Khodja was, of course, as far from being a fatalist as it is possible for any thinker to be, though this was immaterial to his French critics in the 1830s and in the next chapter we will consider in more detail as to how a theory of humanitarianism meshed with the idea of medicine and a European historical fatalism to create a world in which the claims of Khodja could be denied on the basis of his cultural heritage; proven by the pigmentation of his skin rather than through a contestation of his ideas.

3
On Humanitarian Desire

3.1 Barbara Leigh Smith Bodichon

Barbara Leigh Smith Bodichon was a classic nineteenth-century figure in that her significance rests on her having inhabited a whole series of different roles during her life. As the title of Hirsch's biography has it, she was a 'Feminist, Artist and Rebel'. In London, she was the leader of the Langham Place group, close to George Eliot, the Rosettis and Gertrude Jekyll, while her interest to historians of gender also lies in her considerable correspondence in which she reflected on political questions and revealed much about the lives of mid-century British women who were striving for greater personal independence.

Her archetypal qualities were found in the manner in which all of her endeavours seemed to rest upon attitudes towards the world which might be described as progressive, humanitarian or enlightened: based on what was believed to be a fundamental beneficence where the good life was achieved as much through a devotion to others as it was through personal gain and development. In commentaries on Barbara Bodichon, the one aspect of her personality which comes across universally is her sincerity, for she formed a part of movements of modern change and her commitment to these emblems of change could be found in an authentic fashion in the practice of her life, as well as in her writing.

Following on from the previous chapter's consideration of the 'idea of medicine', I now wish to explore further the ideas and sentiments which underpinned the colonial medical mission. This will involve challenging the sincerity of thinkers like Barbara Bodichon in examining the manner in which words became deeds. We will look at the ways in which progressive political ideas such as Bodichon's played themselves out in practice, with a rather closer concentration on the character of her good thoughts and ideas than is usually the case (for literatures on nineteenth-century progressivism tend towards poles of adulation and excoriation).

To take just one example, Barbara Bodichon's husband, Eugène, was a leading medical theorist of the Algerian colony. Much of his writing advanced ideas about race and extermination which, as we shall see, had a malign influence on the lives of Algerians. Yet he was also a devoted medical practitioner who risked his life to care for the poor of Algiers. In broad terms, modern literatures see evidence here of the goodness and badness of Bodichon's character – or Hirsch's more naïve view of a man 'who combined wide scientific interest and medical skill with complete disinterestedness'[1] – but I tend to think that a more nuanced and complex account of these two stances and their interactions needs to be developed. At the very least, we ought to consider investigating the way in which individuals and their actions might be seen as being both locally benign and structurally malign.

Nevertheless, this is to run on too quickly, for we first need to know how Barbara came to be in Algeria. The answer was that she and her sister had come for medical reasons for they had been suffering from what had been described as a kind of exhaustion in England (there was talk of a failed engagement) and their father believed that the warm Algerian climate would provide a suitable cure. The choice of Algiers as a spa – 'The English have always had a genius for discovering unusual pleasure resorts' as Burton remarked – was somewhat paradoxical given that when they arrived in 1856 'the whole province was fever-ridden, and that every year malaria and cholera took their toll of the unfortunate French colonists; that the houses were often damp, and the air after a high wind so laden with sand and dust that the consumptives spluttered and coughed with added distress.'[2]

From the beginning of her time in Algeria, therefore, Barbara Bodichon was confronted with an unusual and extreme environment. She had come to a place which was believed to have remarkable restorative properties, yet she cannot have failed to notice that for both locals and colons, this was still an extremely dangerous place to live. Such disparities could of course be observed in cities like the English capital from which Bodichon had travelled, and from London she brought with her the impulse to try to raise the aspirations of human societies such that some of the wellness of the rich could be shared by the poor.

It was in these surroundings that she met Eugène Bodichon, an older man and something of a free spirit, who had worked in Algiers for a number of years. He was known as a political radical whose view of the world was strongly influenced by Comte, Louis Blanc and Ledru-Rollin, and his tract *L'Humanité* had been banned for its anti-Bonapartism. Even in a culture of outsiders, Eugène was seen to be an idiosyncratic character and Barbara's biographers have speculated that it was this which proved attractive to her. We do not know how the two met, but Burton guessed that it was through Barbara's brother Ben, a big-game hunter, since Eugène was famed amongst

such men for his skill in patching them together after accidents in the field.[3]
At the Bodichons' house:

> Two pictures used to hang on the walls showing Dr Bodichon gathering
> up such pieces as were left of a certain M. Bombonnel's face, which had
> been appallingly mauled in his encounter with a panther. The list of
> wounds comprised five gashes on the left hand, the animal's teeth hav-
> ing pierced it in three places, eight wounds on the left arm, four in the
> head, ten on the face, four in the mouth, the nose-bone having been
> broken, five teeth wrenched out, and the left cheek below the eyelid torn
> to tatters. Yet so skilfully did the doctor patch and sew the pictures
> together that in a very short space of time M. Bombonnel was again at his
> favourite pastime – panther hunting.[4]

By the 1850s, Algeria was not only developing a reputation as a suitable
sanitary destination for health tourists but also as an amenable locale for
other forms of entertainments. I do not want to extend this next thought
too far but we shall go on to see that hunting was also a preferred means of
describing the pursuit of local human quarry, who were in turn frequently
animalised in European texts. One might also remark that the vivid description of Bombonnel's wounds is quite unlike any description I have seen of
those who were rather less visible in French texts of the nineteenth century:
Algerian corpses and bodies injured in battle, or indeed the Algerian sick
tended to by French doctors.

When Barbara first arrived in Algeria, she saw local society in familiar
textual terms as a 'playing out of the Arabian nights' – 'We pass some
Sinbads'! – in which Algerians evidently took on the role of amusement and
diversion that at home had been performed by the characters in such tales.[5]
Paraphrasing Barbara, Burton wrote that in Algiers 'Beauty walked hand-in-
hand with filth, and native grace with the most barbarous superstition.'[6]
Such a picture came, though, not just from the look of the city, for her view
of what she called 'this extraordinarily lovely and weird country' also drew
on a political analysis, in which her feminism led her to 'regard the position
of women in a given society as a useful yardstick for judging its "maturity".
By this measure, she regarded any Mohammedan society as more "barbarous" than any Christian one'.[7] While feminism might notionally depend
on a form of universal solidarity, it was more realistic to see that it particu-
larised on the basis of the historical status of its subjects.

Such a view was of course based on the most cursory of glimpses of a single city of a place whose peoples she did not know and whose languages she
did not speak, but Barbara Bodichon was typical in not allowing such things
to impede her ability to seem to comprehend and describe the culture in
which she found herself. This desire to depict and subsequently, quite logically, to improve and remodel Algerian culture, found its expression in a whole

number of ways, from Barbara's many letters home, her art in which she operated in a typical Orientalist mode, her collecting of local pottery which was sent back to the Victoria and Albert Museum and the assistance which she provided to her husband in his own medical mission. Hirsch notes that what must have been her instinctive enthusiasm for her husband's work came partly from her own background, in which she had described herself as a 'sanitarian'.[8] Whether Barbara's similarly minded cousin, Florence Nightingale, would have agreed with Eugène's ideas on the social implications of that brand of thinking about health was moot. As an English house-guest reported, the doctor 'had a serious theory for improving the world in the shortest possible time by the painless extinction of all useless human beings. He would have juries, including a large proportion of men of science, to decide on the fitness of this person or that to live'.[9]

Since existing literatures have tended to view Barbara Bodichon in terms of her contribution to positive, progressive change, one finds their judgements on her work such as Hirsch's that while 'perhaps her landscapes, celebrating the beauty of Algeria, could be seen as complicit in the production of an Orientalist Other for the consumption of Western purchasers...this was certainly not her intention'.[10] It is such sentiments which I wish to interrogate in this chapter, in terms of both Bodichon's Orientalism and Hirsch's assuredness that critics ought to focus on Barbara's good intentions, rather than on their consequences, whose dubious goodness is alluded to here, or connections between intentions and outcomes. A distinctive quality of the history of health, as opposed to the history of medicine, is that it is as interested in results as it is in intentions, and this is of especial import in a place such as Algeria where so many theorists of medicine were so blind as to the effects of their hypotheses and their consequent place within broader structures of thought and of politics. It is in a way ironic that it was multi-talented figures such as Barbara Bodichon, deeply involved in forms of progressive politics which claimed to be able to read societies in new ways which combined the social, the political and the intellectual, who in fact proved so utterly unable to see the character of the colonial society her clan was erecting in Algeria.

We will need to return to the question of whether such a view of the naïveté of the Bodichons is sufficiently critical of their actions, but let us close this section by returning to the theme of health tourism. We have seen that as a patient in the colony, Barbara was something of a pioneer so it should not surprise us that she was also the author of the first book-length study of Algiers as a sanitary destination: *Algeria Considered as a Winter Residence for the English*, issued as part of the *English Women's Journal* in 1858. An advertisement at the end of her book promoted the virtues of an 'English Boarding House' run by a Mr & Mrs Thurgar, 'For the reception of invalids and travellers', announcing that 'The climate is remarkable for its restorative influences.'[11] The notice also announced that a list of 30 'eminent English and French physicians have kindly permitted their names to be used as referees, and as strongly recommending Algiers in cases of weak and diseased lungs'.[12] Most of the doctors listed were from London, but they also included physicians from Paris, Algiers (the familiar names of Bodichon's husband, A. Bertherand and Henri Foley), Cambridge and Norwich. Other early travel guides which praised the ameliorative effects of the Algerian climate included Lady Herbert's *A Search after Sunshine: Algeria in 1871* and C.S. Vereker's *Scenes in the Sunny South, including the Atlas Mountains and the Oases of the Sahara in Algeria*.

By the time that George W. Harris's *"The" Practical Guide to Algiers* was published in 1890 it was clear that the colony was successfully selling itself on the basis of its healthiness, and as a text Algeria was subject to reinscriptions based very much on earlier templates developed by writers such as the Bodichons. Advertisements revealed the need to reassure English tourists of the adequacy of sanitary facilities: the Grand Hotel Continental offered 'every modern comfort and the best sanitary arrangements',[13] the Hotel Kirsch was 'recommended to English and American visitors for its most healthy and convenient situation',[14] while the Splendide Hotel benefited from both 'Natural drainage' and 'Jennings' perfect sanitary arrangements.'[15] The Hammam R'ihra winter resort and hot mineral baths, which was 'at four hours from Algiers on the Oran line', was 'Most admirably adapted for the treatment of pulmonary diseases, marvellous for Gout, rheumatism, etc.' 'Pigeon shooting, wild boar hunt, and all kinds of shooting' were also offered at moderate prices, which would have pleased Barbara's brother. The hotel did however feel that 'It is necessary to put visitors on their guard against misrepresentations and false reports by interested or prejudiced persons, both as to the climate of Hammam R'ihra and the accommodation to be obtained at the establishment, which rivals any to be met with in Algeria', which is suggestive of the manner in which healthiness had become the arbiter of competition in the tourism market.[16] This is further seen in E.A. Reynolds *Mediterranean Winter Resorts: A Practical Handbook to the Principal Health and Pleasure Resorts on the Shores of the Mediterranean*, which offered its readers an assessment of the strengths and weaknesses of different resorts and their accommodation. The very existence of Algerian health tourism, and its accompanying literatures, therefore acted as a form of proof of the success of French imperialism and its historic mission of reviving the well-being of the southern Mediterranean.

3.2   Humanitarianism

When Barbara Leigh Smith married Eugène Bodichon in 1857, it was a true union of progressives, yet when we speak of the mid-nineteenth-century humanitarian ideas which informed their work, from where did these ideas come and how did Barbara's feminism combine with her husband's medical progressivism? Perhaps the first thing to say is that in the early colony – and much to its later detriment – nearly all forms of 'modern' thought were believed to be compatible with each other, and as a consequence we tend to confront *mélanges* of ideas which now seem to contain rather more potential for contradiction than was thought to be the case at the time.[17] In thinking about the 'Algerian complex' – by which I mean here to refer to the conjunction of this set of ideas – we also find some form of explanation of how confused were French actions in the early colony, and how remarkably unsystematic many were in analysing the relationship between ideas and policies. Those rare Algerian voices which survive, such as Hamdan Khodja in the 1830s and the first doctors working in the colonial medical service in the 1870s, had no such problem in developing ethical critiques of the gaps, consequences and duplicities inherent in French humanitarianism.

Writing in the earliest years of the colonial regime, we have seen that Khodja was well aware that humanitarian rhetoric concealed and masked a system which was both in the process of dismantling existing systems of care and which stressed the overriding determinacy of race in a manner which was alien to Algerian society. French claims that their actions were 'based on principles of morality and civilisation' were given the lie by the venal manner in which the *wakf* and other charitable funds which had traditionally served as the focus of welfare in Algiers had simply been confiscated by the invaders.[18] Yet, as Khodja observed, at that very moment, the French were 'giving millions to the Greeks and the Poles! ... aiding these peoples with Algerian gold!'[19] Khodja was evidently alive to the fact that in making these calculations the French were deploying logics about the values of certain kinds of humans and races because they operated with a highly racialised view of the world in which they simply could not perceive the worth of peoples such as those they found in Algeria. A perfect example of this hypocrisy was the manner in which Clauzel had ordered the demolition of the el-Kaïsserie bazaar which specialised in the sale of books and scientific instruments. In such instances, could there not be admitted to be some strangeness to the fact that, as Khodja was to remark, 'with the stated aim of introducing civilisation into Africa, France destroyed a place from which all kinds of learning and knowledge emanated'?[20]

The aim of this chapter is to explore such gaps, consequences and duplicities in two stages: by first looking at universalist ideas which notionally precluded such *lacunae* and then specifically to identify malign forms of particularism which lay embedded in the character of colonial notions of humanitarianism. In doing this, it ought to be recalled that we are dealing here with but one part of a complex of thought about Algeria, other strands of which are considered elsewhere in this book. We also ought not to forget the increase in terrible local suffering which was paired with the development of putatively universalist systems of care, the specific victims of which will be addressed in the following four chapters.

While this chapter will draw on the exemplary work of Patricia Lorcin, it contests an assumption made in most writing about Algeria, which is the notion that French views on race in the colony changed over time (the classic expression of this claim is that there existed a move from assimilation to association). My belief is that evidence from the nineteenth century shows that such shifts were of little consequence as compared with the strength of racial templates developed early in the life of the colony. In concentrating on doctors' writings about race I am seeking not only to show how medical professionals shaped racial archetypes, but also to assert that ideas about race were seen as emanating from medical, scientific theories. A commonly held contemporary view is that medical writing about race can essentially be seen as ethnographic or as belonging to some other branch of the nascent social sciences, but the point about the work of Bodichon, Bertherand et al. was that they and their readers saw their ideas as being as medically and scientifically valid as the discoveries of diseases and vaccines.

As we saw in Chapter 1, recent work by Cooter et al. has identified a strong correlation between the development of humanitarian ideas in 1870s France and the idea that medicine served an important role in nation-making, particularly in times of war. My suggestion there was that this connection had in fact developed decades earlier in Algeria, almost certainly providing a template for the metropolitan trend identified by these writers. In Algeria, after all, there had been no need to sugar-coat the 'particularistic national interests' in which medicine served the interests of the state, for in an important sense medicine had been assigned the role of making an Algerian state and nation.[21]

The analysis of medical humanitarianism is in one way a subset of broader critiques of humanism, many of which also depend on colonial examples of the contradictions which easily flowed from claims by Europeans that they represented the interests of all humanity. We are concerned here, though, with a very local set of circumstances. While it is true that a notional future of common alignment on a civilisational scale informed French thought in Algeria, it was the construction of logics to justify the taking of brutal choices in support of this larger goal which will interest us most. In terms of medical practices, what I think we find in the colony is the development of an ethical archetype, where the assumed beneficence of outcome, which was enshrined in the humanitarian intention, was presumed to hold moral sway over any objections which might be raised on the basis of the actuality of medical and demographic outcomes. Future gains, in the light of the need to assure the progress of a universal history, invariably trumped present realities.

Why though do we find such a stress on doing good in the writings of both Bodichons and is there any sense that the certainty of this intended beneficence in fact masked worries about the ethical import of their ideas as they were translated into practices? In a sense, the proposed goodness of their work was also underpinned by a concern to do justice to Algerians – in terms, as we have seen, of accelerating their linear civilisational advancement towards, on the one hand, modern medicine, and, on the other, feminism – but did either figure carry with them any conception that notions of non-maleficence or autonomy might in any way compete with the framework through which they proposed to reform Algerian society? We will see that such questions lead us to some rather tangled answers, which the Bodichons did not see or chose not to extricate themselves from, such as the question of the relationship between race and justice. If it truly was the case that their work was driven by a sense of universal justice built on the idea of the branches of humanity finding themselves at different places on a sliding scale of civilisational progress, then how could some groups be judged not only to be backward but also deficient in the sense of being undeserving of universal justice, or indeed as blocks to such justice?

Medicine and feminism were both ideas subsumed within the broader notion of civilisational progress which had an incredible power for enveloping and synthesising sets of ideas which might otherwise seem to have had the potential to come into conflict with each other.[22] Colonialism became a broad church in which the civilisational imperative served as both rationale and representation of France's rights. By this I mean that if we ask of our sources 'On what grounds was medicalization justified?' our answer is 'On the basis of civilizational progress', whilst if we ask how such civilisational underpinning was itself rationalised, the response was that the goodness of medicine and education proved its underlying value. In other words, a system of self-serving circularity lay at the heart of the French mission in Algeria.

In exploring what I shall call the failed utility of an uninflected humanitarianism, our view of medicine in Algeria becomes a somewhat familiar one, for the Foucauldian and post-colonial drives of the modern history of colonial medicine have consistently targeted the exploration of gaps between aspirations and realities in imperial settings. Such approaches also owe much to the work of Walter Benjamin. His critique of Saint-Simonian thought is of critical importance in the Algerian case since a great many doctors and other officials conceived of the colony as a form of *tabula rasa* on which Saint-Simonian experiments in new styles of living might be effected.

The critique of Saint-Simonian thought might also be used as a means of designating a more general synthetic trend in French culture in the 1830–50s, by which I mean to indicate that striking tendency for agreement to be found on the value of modern change from right across what now seems to us a series of divergent points on a political spectrum. It is imperative to see that the formation of Algeria was very much a product of that particular moment in the history of ideas.

It was a time, as Benjamin noted, of remarkable convergences of opinion forming around ideas such as the state incubation of a national rail network, with socialists such as Blanc (who saw rail as a first step to a command economy) and right-wing nationalists (who feared being left behind by other countries) and 'liberals' (who wished the state to risk the initial capital that might create a market in rail services that they might later exploit) forming pragmatic alliances. The colonisation of Algeria provided a comparable example for, as Le Cour Grandmaison observed, the value of this new holding was espoused by men seemingly as politically far apart as Louis Blanc and Bugeaud.[23]

Saint-Simonian thought provided a frame and a claimed theoretical rationale for such synthetic projects in which the desire of all parties to identify their own political stance with the national interest led them to support projects of progressive change with remarkable unanimity. Of course ventures like the development of the railway network and the retention of the Algerian colony were not without their critics, but Benjamin sought to show how compelling a case could be developed on the basis of the breadth of support offered for such ideas and how Saint-Simonians and others swept up these separate logics into supposedly coherent holistic rationales which were welcomed by centres of power in government, the army and industry.

Yet, as Benjamin observed, in their assuredness of their completeness and their expressiveness of harmony, what such schema tended to ignore, or seemed to forget, were the most basic of objections to their own theses. In the case of the railways, this related to questions of ownership, corruption, efficiency and safety, which would all emerge in late nineteenth-century France as scandals and points of genuine political difference; though too late in the sense that the basic architecture of railway development had already been established. In the Algerian instance, the possibility that the beneficent colonial enterprise might in any sense not be a good thing could not be countenanced, and the unanimity that could at least cohere around this basic idea (allowing that some might have different conceptions as to whether the good might, say, be quicker reached through a military or a civilian colony) excluded the possibility that others, such as autonomous indigènes, might not perceive colonialism in this beneficent light.

Returning to the theme of the history of ideas in France, let us consider Gill's claim that there was a dichotomisation of political thought in the nineteenth century:

At the heart of the political controversies of nineteenth-century France was a tension between opposite notions of history and time. One viewed history as a narrative of progress, the other as decline and decay. Throughout the period, these conceptions coexisted uneasily, and informed the world-views of creators and thinkers of diametrically opposed mentalities. They underpinned two conflicting notions of the nation, and of the legitimacy of the state: on the whole, republicans had faith in progress, whilst conservatives and monarchist yearned for a lost traditional social order. Aesthetically, the latter sensitivity thrived on nostalgia, and the portrayal of an elegiac or primitivist lost paradise.[24]

How, we might ask, did such ideas play out in a place like Algeria? Was the colony exclusively the domain of republican progressives who saw it as potentially perfect in the manner in which it could serve as a validation of their worldview, or did there also exist the possibility that an Algerian state might be founded on the kinds of values espoused by conservatives? The truth, I suspect, was that the Algerian experiment became something of a confusion of these two notions, for while it became a natural home for all kinds of progressives (liberals, Marxists, Saint-Simonians), the picture some of these groups developed of the potential of Algeria included a valorisation of the primitive and the possibility of the founding of European communities which escaped the tensions and hierarchies of French cities. This was especially evident in Algeria after 1848, when significant numbers of socialists and other leftist revolutionaries were sentenced by French courts to be transported to Algeria to direct their political fervour towards the establishment of experimental societies away from the mainland and the mainstream of French society. Yet, as Cohen notes, even before this

> The socialists enthusiastically supported the conquest of Algeria in the 1830s, seeing the possibilities of establishing model socialist societies and of expanding into the rest of Africa. In 1831 Philippe Buchez, a Saint-Simonian who later became a 'Christian socialist', welcomed the conquest that had just taken place in Algeria, because it would allow France to dominate the Mediterranean; the new holding could serve as a base for 'direct communication with the interior of Africa.'[25]

Ageron notes that there was near political unanimity with regard to questions of nationalism and colonialism, and cites the socialist Louis Blanc's desire for the total conquest of Algeria by proletarian armies.[26]

It was the Saint-Simonians who perhaps best expressed this mélange of social thought in nineteenth-century France. The consequences of this mixing might be described as a politics of accretion, for little thought was given to the compatibility, or policy consequences, of a politics where idea after idea was added to the mix. The significance of this observation for my argument is that in many ways the confusion wrought by such thinking was applied even more easily to the *tabula rasa* of Algeria than it was to France.

The doctor Auguste Warnier was a classic exemplar of Saint-Simonian practice in Algeria, for he began his career as a military doctor in Algeria, became known as a man 'venerated by the Arabs', invested hugely in land there, was elected prefect of Algiers in 1870 and was chosen as a deputy in 1871. It was in this last incarnation that he was the author of the piece of legislation which facilitated the confiscation (or the forced purchase) of indigenous land by the colons.[27] His career demonstrated both the professional possibilities that were opened up by working in medicine in Algeria and the fact that doctors were seen as central to debates about the nature and future of the colony. As a Saint-Simonian and a doctor he was a high representative of the idea of medicine and of Algeria, and it is somewhat typical of such a figure that the policy consequences of his work was to be of such negative import for Algerians, as we shall see in looking further at Warnier throughout this book.

Returning to Gill's dichotomy, it was also the case that one found, particularly in the ranks of the army, a great number of political conservatives who saw in Algeria the possibility of the construction of an autocratic state headed by military figures, which would nonetheless be driven by a progressive desire to create a powerful 'new France' which would serve as an important agricultural and industrial base for the French economy. In this sense, Algeria found itself in an unusual position with regard to the political ideas of the metropole, in that it was both of France's world of ideas and yet also a canvas in which different interest groups rapidly drew lines across its territory which acted to create a very different kind of culture of politics and ideas than that found in the metropole.

The colony acquired an ambiguous form of identity (later expressed in convoluted policies surrounding its administrative status and the citizenship of its inhabitants) in which it was and was not of the self of the metropole. Its very alterity made it a suitable dumping ground for the social problems and contradictions of French society, yet its difference was of a bracketed variety for it needed to be admitted that the hoped-for success of the Algerian experiment might see it becoming French over time; perhaps the stability of its polity might even flow back towards the tempestuous metropole.

As an outpost of France, both of it and alien to it, Algeria was seen as a laboratory in which the ideas of metropolitan thinkers might be played out in relative safety. It was medicalised in this sense as a sick body on which it was legitimate for France to experiment with various social and political cures, for there seemed little danger that European care would diminish the state of a people already perceived to be living in sickness and squalor. Thus in 1848 Eugène Bodichon was able to proclaim that:

> Algeria must be an asylum open to everyone. Bring us black slaves, freed from the interior, emancipated slaves from America, those who have been expelled from Europe and Asia, the communists, the Phalansterians, the Jesuits, the Trappists etc. etc. All will live together under our laws as

simple citizens. Each will follow their own system of socialism as a means of the conquest and settlement of Algeria.[28]

So while France herself sought to resolve the central social and political questions of her century – regarding governance, secularity and questions of identity – it was seen as quite acceptable to intentionally ensure that such debates lay unresolved in her colony. People were, after all, united on the more profound level of race.

As we have seen, the power of Saint-Simonian thought was stronger in Algeria than in the metropole for the reasons that the colony was viewed as being a safer crucible for such experimentation than the homeland, and because the Mediterranean and the Maghreb formed key parts in the Saint-Simonian picturing and historicising of the world. As Émile Temime showed in his comparative study of the 1830s and the 1930s, the progressivist Mediterraneannism of writers such as Camus, Gabriel Audisio and Jean Amrouche owed a great deal to their Saint-Simonian antecedents a century or so earlier.[29] The Algerian experiment itself built on France's earlier experiences in Egypt, serving as a new opportunity to effect what Morsy called a 'mariage entre l'Orient et l'Occident', both through specific personalities such as Ismayl Urbain (who had converted to Islam in Egypt) and through the historical schema which envisaged the revival of ever closer union across the shores of the Mediterranean.[30]

Morsy rightly identifies what he calls the 'rôle phare des saint-simoniens'[31] in Algeria – well conveying the idea that their apparent moral purity served as an aspiration for all in the colony – but I think he trusted such rhetoric too much when he valorised the kind of marriage envisaged in Algeria. As he wrote, 'Enfantin's conception of modernity... moved beyond a simple opposition between a Muslim Orient and Christian Europe', claiming that this represented a 'brief and rare moment of openness to the other in the history of western thought'.[32] Yet how could this be true when the Algerian project involved the effacement of Algerians on a truly epic scale? Perhaps if one mistakes fascination for openness one can see what Morsy meant, yet in matters of race, culture and medicine, while it is true that the French displayed a need to describe and catalogue their Algerian interlocutors, it is almost impossible to find the voices of Algerians in such texts, let alone instances of Frenchmen learning from Algerians (even in spheres such as medicine where locals had a great deal to show incomers about the local environment). Where can we find in this period, instances of Saint-Simonians who saw the obvious connection between their own proto-socialism and the importance of such ideas in Islam? In matrimonial fantasies of this Mediterranean union, one does not need to guess who played the role of the groom and who the bride.[33]

In fact we might ask here whether Algeria was something of a special case which lies outside Grove's identification of another form of progressivism, a so-called 'Green Imperialism', in which Europeans tended to show an unexpected sensitivity to local environments and the ways in which they were managed, absorbing lessons which they could follow in colonies and which they might also export to the metropole.[34] Given that this willingness to adapt gathered pace through the late eighteenth century, and given that Grove suggests that French medical thinkers and physiocrats played a 'quite disproportionate part' in the evolution of this paradigm, why were colonists in Algeria so disinclined to follow such a trend?[35] Why in fact, whilst being supported by precisely the kinds of scientific and medical thinkers which Grove suggests led elsewhere to the development of forms of conservationism, was the Algerian experience so terribly destructive?

The answer to this question is, I suspect, that while the progressivist complex was usually a fairly harmonious entity, there were instances where one form of progressivism could thwart the possibility of the success of another brand of progressive thought, usually in the name of some form of civilisational prioritisation (let us think here of the analogous case of socialism and feminism). The special status, for France, of Algeria as both a neighbour and a place of utter difference, along with the aspiration to establish a settler colony and the volume of mercantile voices which demanded swift financial returns for the investment in the establishment of a colony, meant that environmentalism was a luxury which could be ill afforded. It would also have entailed some form of dialogue with an indigenous population from whom even progressives did not believe one might learn. Added to this, as we saw in the previous chapter, was the fact that the very status of French science was being staked on the Algerian adventure for the medicalisation of the colony was designed to serve as form of proof of the civilisational narrative. This evidently induced a level of insecurity amongst colonists in the early decades for we know of course that at that point French medicine was remarkably unable to adapt to local circumstances and to fulfil this mission, even amongst the settler population.

Many of the more distant colonies studied by Grove – such as Mauritius – were places of greater local colonial autonomy, where much smaller amounts of national energy were focused on the successful establishment of a New France than was the case in Algeria. The place of the tropics in civilisational history was also negligible as compared with the status of the Mediterranean, which had been assigned a very particular role in the story of progress, both in its past and in the nineteenth-century present. While the islands of the Pacific could be romantically conceived as utopian spaces on which to begin history, there existed the more difficult task of realigning and reorienting history around the Mediterranean. It is indeed hard to imagine a French writer in Algeria valorising its historical alterity in the manner of Sahlins:

> The heretofore obscure histories of remote islands deserve a place alongside the self-contemplation of the European past – or the history of

civilizations – for their own remarkable contributions to an historical understanding. We thus multiply our conceptions of history by the diversity of structures. Suddenly, there are all kinds of new things to consider.[36]

While between 1768 and 1810 Mauritius became a major centre for conservation initiatives 'under the influence of zealous French anti-capitalist physiocrat reformers and their successors',[37] the early Algerian colony was very much a product of its moment in the manner in which its possibilities as both a fund of resources and its potential as a market for French goods were relentlessly investigated.

In some senses, the Saint-Simonians sought a return to the romantic mode of envisioning the colony that Grove identified in the work of figures such as Pierre Poivre, Bernardin de Saint-Pierre and other followers of Rousseau.[38] In Algeria we can see that Saint-Simonians would like to have been able to reconstruct the fantastic archetypes of 'the physical or textual garden and the island'[39] which sustained the development of earlier colonies, but the problem with Algeria was that it was in no sense an island or any form of garden. In the early years of the colony it was a fairly unremittingly bleak place whose landscape, climate and peoples were made only more barren, in literal and figurative senses, by the invading French army which razed much to the ground. While such 'symbolic forms' may have 'offered the possibility of redemption, a realm in which Paradise might be recreated or realised on earth'[40] in places such as Mauritius, the French in Algeria dealt in both brutal realities and what they believed to be a level of knowledge of local society which far exceeded that of Pacific islands, where fantastic notions might more easily be writ atop the blank pages of their histories.

Nonetheless, there was a great desire on the part of Saint-Simonians to return to these earlier, rather more comforting colonial possibilities. The project for the medicalisation of Algeria was a stage which might make such optimistic visions more realisable and the Bodichons conceived of their own return to Eden in their plans for the establishment of eucalyptus forests in Algeria. This project was emblematic of the broader form of denial which characterised French attempts to ameliorate Algerian society through science in the period 1830–70 in that the beneficent claim of such new planting was utterly undermined by the maleficence of the colossal destruction wrought on the Algerian environment by Bugeaud and his army.

Bodichon was in fact subscribing to an important branch of the Algerian meta-narrative which concentrated on environmental change and desertification, brilliantly tracked by Diana K. Davis in her recent work.[41] The claim of this story – that the granary of Rome might be re-established through the application of modern science – was wholly undermined by the fact that desertification, where it existed, was much less prevalent than the French claimed and had been effectively managed locally for centuries, while it was France which had in fact ruptured this local ecology and now grandly announced it had the potential for its rebirth.

Such environmental projects drew extensively on medical writing and explicitly revisited traditions of medical ethics. As Grove remarks:

> Aside from the symbolism of redemption and recreation, the ruling agendas of the botanical garden continued to be medical or therapeutic. Hence the underlying analogue of the garden operated within the established Hippocratic ways of defining the well-being or health of man.[42]

In fact, the Hippocratic revival played a key role in the development of the Algerian complex, along with notions of climatic medicine.

We have already seen that there were great divisions between those such as Rozet who argued that 'the climate of Algeria was as healthy as that of much of Europe', and that consequently 'the population generally enjoy good health', and others who asserted that the Algerian climate posed the greatest possible threat to the success of the colony.[43] Eugène Bodichon contended that colonial acclimatisation to Algeria was intimately connected to the origins of settlers, for 'Those of the blonde race tend to acclimatise only with difficulty in Algeria. They have little capacity to cope with the heat, with forced marches or with outdoor work. Alsatians, Belgians and Prussians and others of Germanic heritage either fell ill or became demoralised in the main.'[44] In the case of such settlers and others who suffered from dysentery, mental illness or complications of the liver, spleen or intestines, the best possible treatment was for them to be returned to Europe.[45] Yet such pathologisation of the Algerian climate along racial lines was not the only means by which such claims might be made, as we see in the *Hygienic Manual of an Algerian Colon*, written by Dr Marcailhou d'Aymeric in 1874, in which class was viewed as the chief determinant of risk in the Algerian climate, 'everywhere the upper classes are able to insulate themselves against the extremes of the climate, whilst the poor suffer the hardships of the climate and the land'.[46]

Such ideas also had a deeper connection with the history of medicine and the revival of Hippocratic medicine in the French empire. As Osborne remarks:

> The Hippocrates resurrected by Littré functioned as the progressive symbol of common sense and resonated with the utilitarian sensibilities of the scientists who worked on the periphery of empire. The ancient physician of Cos had explicitly compared the environments for medical practice in Europe and Asia, paying special attention to physical and moral differences of the inhabitants, and correlating them with the physical nuances of climates and soils. For expeditionary physicians, Hippocrates appeared as a sort of bridge – one of the few – between the

metropolitan world of their medical training and the new world of sand and fevers encountered in Egypt and Algeria.[47]

The renewal of Hippocratic medicine also acted as a neat link in the history of Africa, for in revitalising the learning of the ancients, the French were fulfilling their neo-classical mission of civilising Algeria. Neo-Hippocratic medicine tended to strongly support the expansion of the Algerian colony by 'insisting on the flexibility of European physiology – the so-called cosmopolitanism of the European race – and on the ability to acquire qualities of resistance to morbid climates similar to those perceived in indigenous populations'.[48] Some French doctors in this tradition, such as J.C.M. Boudin, argued that 'exposure to disease in Algeria may have some health benefits for Europeans.'[49] Others, such as Périer, contended that the endemic risks to health in Algeria would only truly be overcome when French settlers had intermarried with locals and produced a generation of settlers who bore within them the resistance of Europe and Africa. Yet such ideas found little influence amongst a settler population which had little interest in meaningful engagement with locals, let alone in marrying and co-founding a new culture. Let us now explore quite why such ideas of *métissage* could never be incorporated within the universal spirit which the French brought to Algeria.

3.3   Race

The central problem with the forms of humanitarianism outlined above was that they were not conceived in a universal fashion in Algeria. They were universal but for the caveat of race. Universalism here was essentially an aspiration that at some future point in time all peoples might join Europeans on the civilisational pathway which had been lost outside countries like France. Given that race was offered as an explanation for such waywardness, and not simply a deficiency that might be overcome in journeying towards a universal future, some rather more serious questions come to mind about the kinds of racial sacrifices that might have needed to have been made to ensure the success of the higher civilisational goal.

Doctors dominated discussions of such subjects in the colony, in part because they had direct access both to a theoretical culture which they had brought with them from France – which allowed them to analyse the traits of races just as much as it did the diagnosis of their civilisational level – and to local peoples whom they met through their clinical work. Chief among such theorists was Eugène Bodichon.

In his work, Bodichon claimed to speak as a scientist and to present a scientific view of the world, but one of the most striking aspects of his *oeuvre* is its peculiarly unscientific character. I speak here not of judging the qualities of Bodichon's work by modern standards but simply of considering the scientific method and Enlightened thought as they had emerged in France in the decades before Bodichon wrote. A defence of Bodichon might contend that his work was founded upon the kinds of observation valorised in the new science, but this was patently not the case for although Bodichon and other writers were deeply influenced by the descriptive culture that was growing up around them, they entered into no deep or comparative studies of Algeria. In fact, Bodichon's 'science' was formed upon a highly romanticised fusion of history and moral enquiry, for ultimately all key decisions were to be made on the basis of a series of ethical judgements about the past.

The Ottoman empire which had ruled in Algeria before the arrival of France needed to be destroyed because it was 'corrupt, immoral and a scourge on peoples'.[50] World historical examples from Asia, Africa, America and Australia proved that it was universally true that desert peoples were 'brigands': this was 'a historic law which we can take from the past'. 'Would', Bodichon then asked, 'this always be the case?' His answer was in the affirmative, unless 'they could be changed through conquest'. Russia was, however, the only country which had succeeded in such – and the choice of words is crucial here – 'humanitarian work'.[51] War was not, therefore, simply a necessity; it was a moral good, and distinctions between civilising and conquering were obviated, as Hamdan Khodja had noted at the moment of conquest.

The particularity of this account of universalism was continued in Bodichon's historical assessment of religion and empire, which showed that France, England and Spain had traditionally 'allowed missionaries too much influence in their empires'. The Dutch were 'much wiser' in not doing so, for history showed that missionaries often tended to 'go native' in the field, even going so far as to prefer locals over their own countrymen. This did not, however, originate in a sense of 'universal justice' but often came from a hatred of colonists who might supplant the place of missions.[52] An important tension surrounding universalism in Bodichon's work emerged here, for it was vital that he found some moral basis for the denial of universal justice. His castigation of the self-interest of missionaries was really, however, a means of masking his broader concern that 'going native' entailed the designations of human equality between Algerians and one's own people.

Bodichon's much earlier *Hygiène Morale* (1851) was in many ways a perfect synthesis of climatic-Hippocratic and racialised ideas of medicine and as such seemed to have been at the cutting edge of scientific medicine in the middle of the nineteenth century. In its huge desire to establish both its innovation and its efficacy, it seems all the stranger to audiences today. Its thesis was summed up in Bodichon's maxim that 'Man is made by his climate and his race.'[53] Three races in particular were of interest to Bodichon:

> The European, who was pleasant and sociable, a theoriser who was ceaseless in his investigation of new questions. ... The Asian, who was anti- social, divided by castes, unchanging, credulous, and attracted by

metaphysical ideals...The African who was hostile, violent and driven more by his instincts than by reflection.[54]

To our eyes, this was evidently ethnography dressed up as medical theory. The Hippocratic addition came in his claim that these three ethnic groups were associated with, respectively, cold, hot and hot/cold climates. In many ways, it was the dual climatic characteristic of Africa which made her and her inhabitants potentially the most dangerous of places for Europeans naturally following their own path of the 'investigation of new questions' in the colony.

According to Bodichon, there was a demonstrable medical history of Africa which showed that it was a place where peoples would rise up through the power of pillage and war (and Bodichon was most certainly not referring to the French here), where passions rose so greatly that 'perversions' such as homosexuality were prone to thrive.[55] It was therefore, of the greatest necessity that 'a severe and puritanical education' system be imposed on Algeria to combat the risks imposed by the climate 'as an agent of corruption.'[56] Humanitarianism was therefore being established as a means of counteracting or combating race, based on the assurance that culture might best nature.

There was, however, some confusion in Bodichon's system for he could not assess whether race or climate was the truer origin of degeneracy. For this reason, European settlers needed to understand the risks they were undertaking in moving to Africa. Bodichon noted that after spending a number of months back in Europe, colons tended to return to Africa as better, less selfish and more reasonable people, which he claimed 'proved that the Algerian climate exercised a debilitating influence on Europeans'.[57] He could not, however, decide whether the mode of transmission of this moral change came through a form of intoxication or through the nervous system. He certainly did not consider that there might have existed different characters of social life in metropolitan France and colonial settler society.

The more Bodichon's work progressed, the more nightmarish the picture he painted became: Algeria made men lazy and torpid, violent and ferocious; it led to a certain 'derangement of intellectual and mental faculties' if one was exposed to the country for a period of years.[58] He went on to explain that 'this derangement was characterised less by a diminution of energy levels, than by a loss of a sense of harmony', for 'the wind of the desert, whenever it gusted with violence, brought with it brawls, murders and suicides'.[59] We have already seen that his theory of suicides in Algeria was an important one in the medical racialisation of the colony.

Although noting that his views were amongst the 'more extreme' to be found in the colony, Lorcin is, I think, absolutely right to note that both the importance of such ideas in framing ideological discussions in the colony and the fact that they came from a man seen as 'something of a philanthropist, who advocated reform for the workers and underprivileged in France'.[60] My suggestion is that we might go beyond Lorcin's note on the coincidence of these phenomena, to suggest that a central structuring reality was at work here, which was that the racialisation of Algerian society – and the consequences which flowed from this – were assured by Bodichon's humanitarianism. In the colonial context, philanthropy and its medical function could serve as subtle forms of warfare that accompanied Bugeaud and the generals' more overt conquest of the Algerian peoples.

If such a bald statement seems somewhat overblown, let me hint now at discussions in Chapter 4, by noting that Bodichon was one of the single most important theorists of 'extermination', who kept such ideas alive in colonial debates across most of the nineteenth century. In insisting on the medical and scientific validity of specific racial categories and ordering in Algeria, Bodichon was essentially an advocate of the idea that categories of the human and the subhuman existed in the colony.

Just as importantly, in Chapter 5 we will begin to look more closely at the connection between race and the scope of the duty of care in the colony. If the division of Algerian society along racial lines could be described as a medical theory, did it not follow – especially at times of pressure and rationing – that doctors owed less of a duty of care to suffering Algerians than they did to Europeans? It may seem odd to us to describe the denial of care to a Berber in favour of retaining scarce resources for the possible care of a Frenchman as a *clinical decision* – for we would see it as having a social motivation – but in the nineteenth-century Algerian context, I would suggest that this was far from anomalous. The status of Algerians as humans, as brothers, and their species rank evidently had a powerful effect on the desire to truly heal individuals – as opposed to the care of notional persons in a generalised medical belief system.

Writing about the origins of Bodichon's thought, Lorcin noted that:

> While Bodichon's preoccupation with anatomical and physiological characteristics can be imputed to his training as a physician, his overriding theme of civilization versus barbarity, his linkage of physical appearance to moral characteristics and his dismissal of the Arabs as less civilizable because of their physiognomy, inherited traits and social organization were all evidence of prevailing ideas on race.[61]

The importance of such a statement, though, is surely that such 'prevailing ideas on race' were seen by Bodichon and his readers as deriving from medical theories. They were emphatically not seen simply as ethnographic generalisations but as medical-scientific observations possessed of the same levels of truthfulness as, for example, Laveran's work. Here I think Lorcin concentrates overmuch on the inherent strangeness of such ideas from our own perspective, when no such sense of estrangement was apparent in their reception in the nineteenth-century colony. Lorcin in fact later acknowledges a greater degree of connection between medical and racial theorisations, noting that 'The contribution of the medical corps was especially significant to the scientific validation of racial concepts as well as to their vulgarization' and that 'The statistical analysis characteristic of their methodology gave their work, for the lay reader especially, that aura of infallibility.'[62]

In one sense, there was nonetheless an anomaly to be found in Bodichon's writing for by the time he was writing *Hygiène Morale*, the great medical fears of the period of the conquest had generally been assuaged and mortality rates amongst Europeans had declined. The structures of the medical state had been in place for some time and were generally viewed with optimism as the means to assuring the health of settlers and winning the hearts of locals. As we have seen, Algeria was even being promoted as a healthy destination for tourists. Yet Bodichon worked in that medical system and we might ask why his vision of medicine in Algeria was so relentlessly bleak. My instinct is that the narrative of medical improvement was actually much more limited than texts would want us to believe, essentially restricted to the containment of mortality rates amongst settlers on the coast, bearing little relation to broader patterns of disease and death.

In Bodichon we see the germ of what would become the ultra-colon mentality, characterised by a profound hatred of locals and the desire to isolate European culture in Algeria from admixture with impure peoples. Ageron and others have noted how prevalent such a mentality became in the 1880s and 1890s. Prochaska's study of the Bône also stressed the coming together of a definable colon culture at that time, where before groups of Europeans (we must remember that most European settlers in Algeria were not French, but primarily from countries such as Italy, Spain and Malta) had lived relatively isolated lives as distinct national communities.[63] Bodichon seemed to foresee such a culture and stand as an early representative of its values, which can also be seen in his notion that it was necessary to

[i]mplant in Algeria a southern [European] population, of the tanned race; since they would acclimatise well to Algeria and stand up to hard work; though such a group needed to be governed by the spirit of the blonde race, the spirit of the north of France.[64]

He was repelled by southern Europeans, for elsewhere he described such 'européo-algériens' as 'those who have fallen from Christian civilization', alleging that Italy and Spain had 'vomited' their 'galley slaves' into Algeria, and France had been little better, sending her 'crooks' and drunks.[65] Such comments chimed with the views of the author of the pamphlet *Should France Retain Algeria?* who argued that the so-called moral conquest was undermined by the tramps and labourers, 'with whom the Bedouins would hardly be flattered to be compared'.[66] From such comments, one can understand why groups within the military and civilian administrations felt superior to colon culture, yet others such as Ernest Mercier called, in 1880, for 'us to become united, without caste distinctions, so as to finish this great national work of conquest through civilisation'. I include this quotation mainly for its pithy demonstration of the indistinguishability of civilisation, the nation and war.[67]

Like all prophets, then, Bodichon was an imperfect guide to the future of the colony: where he saw colon culture continuing to be driven by a caste system between the French elite and the Mediterranean classes, it was eventually to be formed by a mixing of these two groups which had made the paranoid, medicalised, racialised culture he theorised. His mistake it seemed was to misinterpret his belief that, 'The climate of Algeria was conducive to the production of sects ... of anti-fraternité', and to not foresee that it was to be splits between Algeria and the metropole which would lie in contrast to the relative unity of colon culture.[68]

3.4 Race, assimilation and medicine in the later nineteenth century

Race was therefore the dominant motif of French imperial thinking about Algeria from the moment of the conquest. In cultural terms, the French had adopted a variegated, stratified system in which Algerians could be divided into a large number of separate ethnic categories, which in fact often bore little relation to the true ties and binds of identity which existed in the country. This was a method which was later adopted in other French colonies, such as Laos where essential differences were alleged to have existed between indigènes (divided into Annnamites, Laotiens, Cambodgiens and Moïs), *asiatiques étrangers* (Chinese and Japanese), the 'Chetty' of French India and the *métis* or mixed-bloods.[69] Both the Algerian and the Laotian examples give the lie to the contention of the early French historian of colonialism Henri Blet that 'Frenchmen have never adopted racial doctrines affirming the superiority of whites over men of colour.'[70]

The legal corollaries to this 'system' were extremely complex, for there originally existed five legal categories of personhood in Algeria: French citizens, European foreigners, foreign Muslims, indigenous Muslims and indigenous Jews.[71] This was made more complicated by the fact that Muslims were bound by both French and Islamic law and there was confusion regarding the possibility as to how they might gain French citizenship. In theory, the situation was clear for both Algerian Muslims and Jews since French citizenship was open to both should they be willing to renounce their faiths, which almost none would consent to (about seven thousand Algerian Muslims became French citizens in the period 1865–1962).[72] Yet the fact that this special qualification was attached to citizenship in Algeria led to debates about its fairness in both the colony and the metropole; in particular because the question was raised as to what nationality such people held if they were not French citizens. The solution to this problem arrived at by the Court of Algiers and the Cour de cassation was that while Algerian Jews and Muslims were not French citizens, they were French subjects.[73] As we shall soon see, the Crémieux decrees of 1870 were to fundamentally change all Algerian thinking about race.

Such legal argument was of course founded on broader conceptual discussions of race and assimilation. As I have already said, the Algerian situation does not seem to fit the model proposed by Betts and others whereby the period 1890–1914 saw the growth of 'scientific racism' and a subsequent discrediting of assimilation and a valorisation of 'association' (which was viewed as a more realistic policy option given the fundamental incapacity of races to mix together).[74] Looking at Algeria, Lorcin argued that:

> By the end of the century, whatever side of the fence one was on, assimilation of the indigenous population was being acknowledged as a failure. On one side were those who had no desire to assimilate them, claiming that they were incapable of being assimilated; on the other were those sensitive to indigenous problems who realised that assimilation to date, far from bestowing the benefits of civilization, had been a disaster, causing the indigenous population to lose out in almost every domain. Association became the more attractive proposition.[75]

While most of what Lorcin says is correct, I think she is wrong to begin from the premise that the assimilationist idea was superseded in Algeria, for it is arguable that there was never a particularly strong assimilationist ideal in French colonial thought in Algeria. Even the idea of medicine as a progressive project was to be an imposition on Algerian society, which served French strategic interests and which reinforced a racial hierarchisation of Algerian society. When one looks at very specific areas of medical policy where assimilation might have been practised, such as the training of Algerians as health professionals, it was not until the 1870s that a school of medicine was established at the University of Algiers and the number of Algerians it graduated numbered in the tens in both the nineteenth and twentieth centuries.

At a regional level, Prochaska's study of Bône shows that while different communities became more and more socially and spatially segregated over time, 'settler intransigence' had 'stymied' assmilationist policies on a consistent basis.[76] In fact, one of the most interesting aspects of his work is his tracking of the development of the Jeune Algérien movement in the 1900s and 1910s, which was made up of educated and increasingly economically successful Algerians who wanted the project of assimilation to be realised in the colony, with the support of French *indigènophiles*.[77] It was ultimately to be the failure of such groups to secure representation for Algerians through legalistic, assimilationist argument which led, in the 1930s, to the formation of Islamist movements such as Messali Hadj's Mouvement pour le Triomphe des Libertées Démographiques.[78]

Medical texts evidently played a critical role in the racial segmentation of Algerian society, particularly since French doctors stationed in Algeria saw themselves as scientists whose work was able to connect scientific understandings of the body with sociological, anthropological and moral judgements which we would now see as lying outside the domain of medical professionals. In Algeria, we see little sign neither of the traditions of anti-slavery that had existed in the French medical profession nor, understandably, the anti-colonialism of doctors such as Charles Frébault, Alfred Giard and Jean Turigny, who had spoken out against imperialism from their position as deputies in the Chamber.[79]

Perhaps the most important racial distinction sustained by doctors was the so-called 'Kabyle myth', in which the Kabyles (a term which was stretched to include all Berbers and inhabitants of mountainous territory in Algeria) were argued to be racially and culturally distinct from Algerian Arabs (Muslim tribes), 'Moors' and Jews. The Kabyles were valorised by the French on the basis that they would be more susceptible to the offer of European civilisation, given that they supposedly led sedentary lives and were not religiously inclined in the manner of the Arabs.[80]

It should not seem surprising that Bodichon was an arch-theorist of the supposed differences between the two groups. In his *Considérations sur l'Algérie* of 1845, he had argued that Arabs were destined to pass on their negative social characteristics in a hereditary fashion – 'A lack of cross-breeding with other races meant that their love of thieving and raping, traits which had characterised their ancestors, had developed and been passed down the generations' – while the faces of the Kabyles revealed their greater civilisation.[81] In fact, 'The best analogy that could be made was to the dromedary. Morally and physically there were similarities between the two. Both originated in the desert, had long legs and necks, large feet and hairy skin, and admirably resisted fatigue and prolonged deprivations.'[82] It was France's duty in fact to foster antipathy between the two races and to set them against one another.[83] Writing almost forty years later, Bodichon was to continue in this vein, performing his familiar manoeuvre of combining history and morality as a claimed science, with his suggestion that Berbers were 'morally superior' to 'all the other races of Africa' because they were 'essentially descended from ancient Europeans'.[84]

The place of Jews in this racialised system was especially testing, particularly following the enactment of the Crémieux decrees of 1870 which had 'emancipated' Algerian Jews in offering them French citizenship without the need to renounce their faith. In a sense, this was a continuation of French rhetoric from the 1830s, when France had proudly seen an adjunct of its invasion as being a freeing of North Africa's Jews from the repressive legislation of the Ottomans. This perspective took little account of the long history of the settlement of Jews in Algeria, and the fact that they had been welcomed as refugees from Iberia in the fifteenth and sixteenth centuries, nor the fact that the chief gain for Jews in the French imperial system was that they occupied an intermediate position between the French and other indigenous groups.

The Crémieux decrees were, however, enacted in great haste by the exiled government in Tours at the time of the Prussian invasion in 1870, with many wondering whether they would have become law at a less febrile moment. Crémieux explained that his proposals would 'see the development of the great principles of the French Revolution, to which all of the French people owed so much, and which Jews had a greater duty than others to proclaim'.[85] Such principles evidently did not include universalism for Crémieux said nothing of Algerian Muslims.

The decrees were to be sources of intense political debate, and violence, in both France and Algeria in the period 1871–1900. Looking back from that latter date, Cohen noted that one of the problems with the decrees was that they had been intended as part of a wider programme of reform which had not come about because of regime change in France.[86] Surveying the protests which the decrees had aroused, Cohen enumerated a long list of objections to them, including those who claimed that the National Defence government had not signed the decrees, those who argued that the bill had been passed at an inopportune moment, those who objected to the idea that Jews should be French citizens and those who contended that the laws were the cause of the great Muslim insurrection of 1871.[87]

It was certainly the case that this formalising of the Algerian Jews' status in the racial hierarchy of the colony was incredibly divisive. Writing in 1890, Raoul Bergot, described the decrees as 'one of the greatest and most shameful faults in contemporary history' which was destined to be a source of 'cruel disillusionment for France'.[88] After 1871, Arabs were able to point to Algerian Jews as an enemy, noting that these too were French citizens and that 'it was them who governed' Algeria.[89]

Algerian Jews, whose position in the French-defined hierarchy had always been tenuous, now found themselves under colossal pressure from both Algerian Muslims, from anti-Semites in the metropole and, increasingly from the burgeoning and unifying class of colons who saw themselves as implacably opposed to the idea of any Algerians being offered French citizenship. This climaxed in 1898 when four out of six representatives elected for the chamber of deputies from Algeria were convinced anti-Semites. Yet the Jews of Algeria had no Zola, no J'Accuse and no sustained defence of the manner in which they had been placed in a position which went against the principles of French Republicanism; for the situation of Algerian Muslims was even worse than that of the Jews and the granting of citizenship to Jews had only emphasised how entrenched French hostility towards Muslims and Arabs trumped political principles. The Crémieux decrees were a perfect example of the lack of thought that existed between ideas and policies, and arguably also between the interaction of the politics of the metropole and that of the colony.

3.5   Conclusion

Healthcare was an important public form of demonstration of France's thinking about race in Algeria, for stratification of medical provision was something which one saw every day, as one passed healthcare facilities segregated on racial lines. Seeing such things, Algerian Muslims were well aware that they lay at the bottom of a racial ladder, and that the quality of care offered to French citizens and Jews was a reflection of the end of a dream of universality which had been vigorously promoted by the French in the first decades of empire. Yet this was of course a consequence of the brand of universalism which had originally been promoted by the French, which was not really universalism at all, since what France had created was a system whereby racial differences were acknowledged under a universalist umbrella. This schema was never designed to foster an assimilationist culture, and was arguably always destined to generate long-term inequity once either the will or the funds to hold to this peculiar kind of universalism waned.

Writing in 1899, Albert Hugues summed up the sense of utter human difference which had been established between Europeans and Algerians from 'the very moment of the conquest' with his claim that on the level of morality, law and religion 'no sense of connection could have been found between French and Muslim populations'.[90] Once this had been established as a basis for colonial society, how much easier it would then be to claim that such human difference could be used as the basis of forms of apartheid or, more radically, purificatory extermination.

It is rarely mentioned that whether one is looking at texts about race or humanitarianism from the 1830s or the 1890s, there are remarkably few differences in the views of such texts or the shape of debates on these themes across the nineteenth-century Algeria. Historians instinctively seek that which changes but I cannot see how a case can be developed that thinking about race and the human changed in any meaningful way in the colony. One reason why this was the case was that such discussions were seen as having their origins in immutable scientific and medical theories. It is again important to stress that while such writing now appears to us to be social scientific, at best, France believed the racial structures of her colony to be based upon profound truths about the human condition elicited through the experimental method.

That notion of a 'human condition' was again of a rather different character than we might now expect, for it transpired that humanitarianism was actually a function of history. By this, I mean that in Algeria the most critical distinctions which were made about people and their place in the human family were based upon the historical status of the race to which they belonged. Individuals were utterly determined by race, and races – including Europeans – were destined to act in a fashion befitting the chapter of the historical drama in which they played. Theoretically, the French owed a duty of humanitarian advancement to the backward peoples of Algeria, but their scientific ability to distinguish between the stages at which those people sat in the great civilisational story meant that it had to be admitted that some might be helped more easily than others. Such ideas both derived from medical and scientific thinking and they informed the construction of medical and scientific systems of governance in the colony. For this reason, we read the works of doctors such as Bodichon to try to come to understand how they formed this dual function as cultural producers of theoretical knowledge and practitioners of a racially moralised medicine in the colony.

We should therefore be instinctively suspicious of claims to humanitarian beneficence, just as Taithe and others have been when studying the metropole in the later nineteenth century. Such beneficence was a necessary cloak for the reality of stratified medical systems which assigned varying levels of humanity to different racial groups, which is of huge importance to historians of health, who are as interested in the denial of care as they are in medical practice. A moral reading of such systems cannot simply contrast the closed medical culture of the colony at that time with our contemporary standpoint, for we possess a small number of works by Algerian writers who saw the social consequences of a French medicalisation of Algerian culture on humanitarian grounds. Khodja understood that a racial imperative drove French policy and that in structural terms imputed care for others was in fact replaced with utter indifference to their lives.

While individual doctors may stand as beacons of unambiguous beneficence, many also tended to want to play roles in the construction of systems which diminished the lives of the very patients who they carefully tended as individuals. Such trends were difficult to see in the nineteenth century in part because humanitarian ideas were bound up in a complex of progressive change which formed a part of a very particular political culture. That culture was characterised by a great optimism, a valorisation of ideas and theories over facts and outcomes, and it was essentially a form of limited resistance to autocracy in France which allowed for the development of new ideas when they were universally agreed upon across the political spectrum. In France, there were distinct limits as to the development of such politics, but in Algeria – whose invasion neatly coincided with this trend – there was much greater potential for experimentation.

For this and other reasons, Algeria came to be seen as a special case. Earlier colonial ventures had also been developed along lines that matched metropolitan political debates – as seen in Grove's Mauritian example – but Algeria was conquered in a more thoroughly scientific way, befitting its status as France's first acquisition of the progressive, modern age. Yet it was the historic past rather than the scientific future which ultimately determined the medical racialisation of Algerians for, unlike Mauritius, Algeria had a vital role to play in the drama of human history. The importance of this role increased the likelihood that France needed to act in radical ways to ensure the difficult task of the temporal reorientation of the Maghreb. In later chapters, we will see how seriously the possibility of the expulsion or extermination of the people of Algeria was countenanced as being perhaps the only means to achieve this historical necessity.

It was conversations, in person and in texts, about narratives, theories and ideas which allowed for a distinctive conjoining of humanitarian and racial ideas in Algeria, which in turn can be seen to have engendered an administration of sickness for Algerians. Thinkers such as Barbara Leigh Bodichon are of interest to us because her career in the colony was characterised not only by conviction but also by desire; and it was this desire to benefit humanity which led to the erasure of the Algerian in her work. In essence, there was no need for an Algerian presence or voice in her thought, because the strength and purity of her desire for all humanity assured the absolute beneficence of her endeavours and that of the progressive project in general. It is arguable, I believe, that the disappearance of the indigène from this vision was a vital twin to the more literal description of the expurgation of Algerians which we find in the writings of soldiers who fought Maghrebin foes, for both authoritarian and progressive politics pointed to the possibility that the Algerian problem might be solved without reference to the existence of local peoples.

4
On Extermination

4.1 'Et on y met le feu'

On 17 June 1845, Colonel Jean Jacques Pélissier was faced with a choice. As part of Bugeaud's push to pacify the Kabyle region, Pélissier had been assigned the task of subduing the Ouled Riah tribe. This was not his sole mission, but it was an operational objective which needed to be accomplished before he could set about his other campaigning tasks that formed a part of the broader extension of the Algerian territories.

Pélissier's choice related to how he would deal with the Ouled Riah, who had fled from their villages in the foothills up into the higher mountains, where they had ensconced themselves in a complex of caves. Such tactics on the part of Algerians were familiar to Pélissier for they had been adopted by many groups from the Kabyle when faced by French troops. The Kabyles stood little chance of fighting off the French on open ground so – just as they had done under Ottoman rule – they turned to their caves as a last redoubt.

On this day a number of casualties had been sustained on both sides as Pélissier had sent scouts towards the caves, some of whom had been shot, and it was assumed that the tribe had also sustained casualties for Pélissier's men had fired into the caves in support of their *zouaves*. The dilemma, therefore, that Pélissier faced was as to how he could end this stand-off so as to reach his ultimate goal, which was to effect some kind of peace treaty with the Ouled Riah so that he might then move towards his next strategic objective.

Believing that the option of sending scouts and negotiators had been effectively annulled by the aggression of his foes, Pélissier perceived himself as having two starkly different options: he could wait and effect some kind of siege or he could use force to realise a quicker resolution to this deeply frustrating situation. He opted for the latter path and, given the inherent danger of sending his men into the dark caves, where they would make fine targets for those hidden therein, he decided to smoke the Ouled Riah from their hideout. As one of his scouts remarked, 'Et on y met le feu'.[1]

An order was given that flaming bundles of twigs should be thrown into various parts of the cave complex and Pélissier felt sure that he had arrived at a means of ending the resistance of the Ouled Riah in the caverns at Dahra. Over a course of a number of hours, in which a further series of burning bundles were launched into the caves, Pélissier was to be proved right.

At one point a series of shots rung out in the caves, which led Pélissier to believe that some of the Ouled Riah wished to escape the caverns, whilst others exercised lethal force as a means of ensuring loyalty to the commanders of the tribe. After 12 hours the flames and the smoke had begun to disperse, so at this point Pélissier sent a group of his men into the caves with instructions to exercise great care in assessing whether the Ouled Riah were indeed subdued and in a position to now pact with the French army.

What those soldiers who advanced found in the cave were around six hundred villagers. Almost all of them appeared to be dead, but after dragging the bodies from the smoky caves into clearer air, it became clear that perhaps fifty or a hundred of the tribe had survived, able now to make peace with the French army.

4.2   If not to heal, then to kill

It is legitimate to ask quite what the massacre at Dahra has to do with a history of medicine in nineteenth-century Algeria. My response to that question is to assert that Dahra is of crucial import in helping us to understand why histories of health can lead us to new understandings of the past. By a history of health, I have indicated that I mean an investigation which looks at questions of dying and quality of life as well as at medical care; which stretches its cast beyond medical professionals to look at the ways in which the actions of groups such as soldiers and administrators impacted on the well-being of Algerians. In following Fanon we might say that such an enterprise also concerns itself with the effect of such actions on the French actors themselves, intimating also that a historicisation of Fanon's ideas refers to the establishment of tropes of behaviour which structured the lives of Algerians for more than a hundred years.

In the simplest of terms, France brought death as well as healing to Algeria, and I should like to assert that death and healing were rather more intimately connected than we might usually assume. Dahra, I believe, shows us that French colonialism in Algeria was founded upon an exterminatory logic in which a rather blunt equation structured many of the colonist's dealings with her new subjects: either Algerians accepted France and her beneficent offer or they rejected her on the understanding that a logical consequence of that rejection was death. This in turn led to a sense of generalised frustration that Algerians did not seem to understand that they faced terrible consequences if they failed to accept France's gift.

We tend to see, or we tend to want to see, 'healing' and 'killing' as forms of opposites, but if we think about the way that ethical codes are enacted in practice then the journey between care and killing is much shorter than that which we picture along a moral spectrum (part of the problem here may be with the idea of picturing morality). A logic of annihilation may easily follow as a direct consequence of a failure of a policy of healing, and we might admit there to be connections between the way in which the human is viewed in such descriptions of sanitation and cleansing.

This strategy, and the moral arguments on which it was underpinned, are discussed remarkably little in literatures on Algeria, so what I want to do in this chapter is to reveal the hitherto underestimated place which exterminatory policies played within the decision-making environment of the nineteenth-century colony, using Dahra as a means of exploring in some detail the ethical nuances which structured such ideas as they were put into practice. We might also note that there has been no detailed, modern study of Dahra and its reception in metropolitan and colonial literatures. These discussions then lead into later chapters' assessments of other critical areas of the history of health in nineteenth-century Algeria, such as the role which French medicine and administration played in famines, and the question as to whether or not there was an Algerian genocide. What I shall seek to argue here is that in the case of both the violence of indifference which we find in famines and in the planned killing of genocide, eliminationist literatures played a crucial enabling role in forming a broad culture of attitudes towards death and the value of life in the colony.

That culture needs to be reconstructed by looking closely at the language and ideas of sources from the nineteenth century. My concentration on published work such as memoirs reflects my conviction that in coming to understand the moral world of eliminationist culture, we need to see how such concepts were discussed in public, in part to see what might be openly remarked upon at that time. Was it the case that the relationship between the French and the Algerians was structured not just by the notion of moral imperialism and the idea of medicine, with all their racial and dehumanising hierarchies, but also by a central tension between a desire to both exterminate and heal Algerians? In the conceptualisation of the body of the Algerian population and the specific bodies of individual Algerians, might it have been the case that the medical imperative was in fact an expression of guilt with regard to the knowledge of the slaughter of Algerian culture and peoples?

While doctors and other health professionals may be relatively absent from this chapter, the language of the medicalisation of the colony is very much present, and I hope that it will become clear quite how connected were French doctors' conceptions of their role in healing the Maghreb and the ideas of those soldiers who assigned roles in that medicalising mission. An important feature, therefore, of histories of health is the manner in which they argue against the existence of medical spheres which are in some way distinct, from broader environments of policy-making and implementation.

This chapter evidently forms a part of this book's broader task of the recuperation of Algerians into the history of Algeria in the nineteenth century, and in doing so it also works against prevailing views regarding colonial policies towards Algerians. Where there has generally been a desire to view such politics as moving between ideas of assimilation and association, I assert that an annihilatory instinct has tended to be lost in the move from letters, political debates and memoirs of the nineteenth century (where it is much discussed) to discussions in the metropole and in later histories, which ignore one of the dominant strands of thinking about Algerians in the colony. I call, therefore, for a reassessment of the *razzia*, critical texts such as the works of Desjobert, and for a deeper understanding of the relationship between desire, death, guilt, healing and extermination in nineteenth-century Algeria.

4.3   Ethics

Colonel Pélissier was a soldier, but he was also a representative of a civilising mission. What ethical questions lie at the core of his decision to asphyxiate 500 men, women and children at Dahra in 1845?

Such a question is not one which is simply put historically after the fact, for it relates very directly to debates which took place about Dahra in both Algeria and France in 1845 and in the following years, for Pélissier's actions at Dahra were seen to resonate more widely and to expose key questions about the ethics of empire. In both the Chamber of Deputies and the press, we read of a metropolitan revulsion at Pélissier's actions, which was extended by some to claim that the fact of the massacre at Dahra was more broadly representative of the character of France's unjust conquest of Algeria. Yet in the Chamber, the press and, as time went on, in memoirs from the colony, we also find strong defences of Pélissier's actions which are of interest to us not only for their arguments but also for the manner in which they too cite the ethics of Dahra as being at the heart of morality of empire. Both supporters and opponents of Pélissier and his commanding officer, Bugeaud, argued that the importance of Dahra went beyond the event itself for it expressed the very essence of the Algerian project: it could be studied as a set of actions and a set of decisions informed by behaviours and moral judgements which encapsulated what France hoped to achieve in Algeria.

One reason why the massacre at Dahra lent itself so well to such ethical readings was that all sides chose to see the events of June 1845 as being centred around a series of very clear choices and decisions which were explicitly underpinned by moral stances. In this sense Dahra acquired an historical neatness in which it could then perform its role as an exemplar in moral debates about Algeria, though one of the things we will see when we look rather more closely at the events of that day is that the contemporary rush to find symbolic import in these events sometimes occluded the complexity, confusion, uncertainty and obfuscation which actually surrounded the decisions made by Pélissier and his foes.

The question of the agency or otherwise of the Ouled Riah and the choices which they made in the series of events which were to lead to their destruction are of as great an importance to us as those made by French troops, not least because supporters of Pélissier relentlessly stressed the immorality of the decisions which had been taken by the leaders of the tribe. For the historian, this approach poses real difficulties, because while we possess French sources which speak of the terrible injustice done to the Ouled Riah, we know of no voices from contemporary Algerian sources which speak of locals' reactions to the events at Dahra, not least, of course, because most of the potential testimonies of Algerian witnesses were destroyed in the massacre. Nevertheless, it is important to see that our sources, particularly those which support Pélissier, saw an ethical dance being played at Dahra, in which the actions of one party induced a very particular, almost programmed, response, in the other, as this dance moved sequentially towards its logical end point. Dahra was not therefore simply expressive of a set of moral choices made by the French and by Algerians, but of the moral character of the kinds of conversations which they had – evinced especially starkly in this situation where both sides were placed under such great pressure. Pélissier did not therefore come to the hard decisions which he took at Dahra in some kind of moral vacuum, but in a series of responses to both circumstances and the decisions made by his interlocutor.

The logic of all such ethical discussion is towards the exterminatory *either/ or* which I am suggesting structured French encounters with Algerians. Either Algerians in general accepted the goodness of the French offer of healing and civilisation, or they faced the merciless consequences of this rejection of a beneficent offer. Either the Ouled Riah accepted the goodness and justice of Colonel Pélissier or they accepted the annihilatory consequences of that repudiation. The maleficence, the doing harm, of Pélissier was of a second order in a schema in which a particular mode of structuring dialogues between the French and Algerians was of ultimate significance. The consequences of actions had moral import, but that significance was subsumed by a set of higher priorities which were capable of switching the seeming ethical status of actions. While a *prima facie* case could be made against killing, higher logics could be successfully appealed to in the kinds of complex, modern ethical systems one found in places such as Algeria. Politics, warring and governance were structured in a fashion very similar to emerging ideas of *triage* in military, and subsequently civilian, medicine, in which hard and definite choices were made whereby acceptable losses were tolerated, bracketed and quickly forgotten, in the name of realistic, achievable ends. Choices were made, but once they had been made those paths that were rejected were not dwelt upon.

Pélissier's supporters could simply have deployed an older style of thinking in which all killing was justified in war – and they might have linked this to the ethics underlying Islamic distinctions between actions in the *dar al harb*, the place of war and the *dar al Islam*, the place of Islam, or peace – but such arguments are only one of the types which were used to defend Pélissier. Indeed, the question of whether the ethics of war overruled the ethics of beneficent subjugation at Dahra was a complex one for there also existed the comparative dimension that was made available by the Ottoman example. While many French writers made much of the moral superiority of French dominance as compared with Ottoman rule, with the implied betterment of Algerians' lives after 1830, this was hard to establish in the case of Dahra, for it was well known that the Ottomans had not resorted to measures such as those chosen by Pélissier when they had trapped tribes such as the Ouled Riah in caves.

4.4   The history of extermination

Why is it that ideas of extermination and genocide play such a small role in historical accounts of nineteenth-century Algeria? Outside of the hagiographic memoirs of colons across the nineteenth and twentieth centuries, it is after all certainly not the case that historians have flattered French imperialism and writers such as Ageron, Julien and Rey-Goldzeiguer have been deeply critical of the gaps which existed between France's claims as to her aspirations in Algeria and the often brutal realities of her rule.

In spite of the acknowledged bloodiness of the Franco-Algerian encounter – most especially in the years 1830–80 and 1954–62 – there has been little sustained consideration of the idea that an exterminatory urge was a structuring reality of the Algerian colony. Lying beyond the possibility that no such urge existed, there are at least five reasons as to why extermination has been little considered. First, there is the possibility that while discussion of extermination may have been noted in French politics, this was essentially seen as a rhetorical strategy rather than a policy that might be enacted. Second, I think there has been an understandable resistance to what seem to be too neat a connection between the undeniable facts of the existence of discussions of extermination and the demographic collapse of the period 1830–80. Relatedly, in twentieth-century writing there is a concern with the need to prove through documentary evidence that genocides were planned enterprises. As we shall see, this is not as difficult to establish as might be imagined in the Algerian case. Third, there is the structural question of some of these issues of war, death and morality being discussed, often interestingly and critically, under other headings such as 'the *razzia*' and 'the massacre'. Fourth, I think a subtext of most historical writing about nineteenth-century Algeria is that it instinctively treats the whole of the period as a time of war, and thus makes certain assumptions about the kinds of behaviours which we find in times of war as compared to those of peace. Finally, and perhaps most importantly, I think most historians display an essential faith in humanity: in that, they want to believe that planned killing is an aberration in human history, rather than a commonplace.

One place which this diminution of the place of extermination does not come from is texts from the nineteenth century, for they are replete with discussions of an annihilatory politics and one of the reasons I stress the term *extermination* is as a means of indicating just how frequently one comes across the terms 'exterminer' and 'extermination' in French texts. Those words do not lie alone, for we shall see that they form a part of a family of ideas that also includes 'massacres', 'refoulements', 'annihilation' and 'razzias', for there existed a great variety of means of discussing the killing or elimination of Algerians *en masse*.

My overarching historiographical view of this situation is that just as twentieth-century French collective memory can be adjudged to be suffering from an 'Algerian syndrome', with regard to the moral and national import of France's role in the Civil War and post-war Algeria, that sense of a syndrome of forgetting and occluding the darkness of the past should also be extended to the nineteenth-century colony. After all, if we think about questions of genocide and empire from the comparative perspective of other settler colonies, it would in some ways be surprising if in Algeria – unlike nineteenth-century Australia and America – we did not find expressions, in words and deeds, of a desire to rid a newly conquered land of its existing inhabitants so that it might be exploited by new arrivals from the metropole. In saying this, I am also hinting at the broader role which an exterminatory politics arguably played in structuring Franco–Algerian relations right across the modern period: a wider scope I shall only be able to allude to in this chapter. It would nonetheless be interesting to study the manner in which certain ideas of life and death were embedded into Algerian history and politics at that very moment when an Algerian nation and polity were coming into being.

The one outstanding exception to this silence regarding extermination came in 2005 with Le Cour Grandmaison's *Coloniser, exterminer: sur la guerre et l'état colonial*. That book's scope is far greater than this chapter in that Le Cour Grandmaison is concerned with establishing that an exterminatory culture underlay all colonialism, though it is important to note that the Algerian example was of critical importance to him in making his case. My hope is that this chapter extends this earlier work, in terms of its scope, detail and ethical focus.

Before moving on it is worth absorbing Le Cour Grandmaison's etymological note that in the eighteenth and nineteenth centuries, the term 'extermination' possessed a considerable range of meanings, ranging from 'the killing of an individual and the dismembering or burning of his body, to summary executions and massacres'.[2] It was this scope of meaning and the potential for ambiguity which, for Le Cour Grandmaison, was partly responsible for the openness with which the term was deployed and the manner in which it was not felt necessary 'to use euphemisms in accounting for that which the term described'.[3] While I think Le Cour Grandmaison is correct to urge caution in the way in which the term 'extermination' was used in the nineteenth century, I would also tend towards a more conspiratorial reading of its ambiguity, in that it would often seem to be the case that writers at that time meant very well to denote mass, organised slaughter when they used the term 'extermination', and that this is confirmed by other material in their texts, though they were well aware that the 'polysemic' qualities of the word provided the protection of some uncertainty of interpretation.

4.5   A realm of brutality

Historians have, I believe, tended to fail to see quite how barbarous a culture existed in the Algerian colony. This may arise from a general scepticism adopted towards colonial rhetoric, in which discussions of elimination and annihilation are accorded the same kinds of doubting as claims that colonialism brought with it civilisation and progress; in other words, that the realities of empire lay somewhere between these two extremes. What I am setting out to do in this chapter is to establish that such a view is not tenable, for there existed not only a culture of elimination in nineteenth-century Algeria but also the personnel – inside and outside the military – to put that vision into practice.

When we return to the debates of 1830s and 1840s, one of the things which most surprises modern sensibilities is the degree of conviction which most writers displayed that the conquest of Algeria needed to be an enterprise of great violence. This was well expressed by Victor Hugo in a more general set of criticisms of French colonialism, where French imperialism was compared unfavourably with the brutality of the British and the Russians:

> France does not understand how to colonise and will always struggle to succeed in this area. A complete civilization like her own is a delicate and thoughtful thing, full of humanity and unconnected with the worlds of the savages. It may seem strange to say this, but perhaps what France is missing in Algeria is the application of a touch of barbarity ['Chose étrange à dire, et bien vraie, pourtant ce qui manque à la France en Alger c'est un peu de barbarie'].[4]

Yet if we think of the ways in which Algeria had traditionally been represented in Europe, it may not 'seem strange to say' that France would need to adopt a barbarous approach to pacifying her new territory. After all, for centuries the *Barbary Coast* had had a reputation in Europe as a place of danger, lawlessness and cruelty, its inhabitants famed for enslaving even white Europeans as well as black Africans. This had only been tempered to some extent by Ottoman rule, for it was understood that the Turks had merely contained Algerian society, rather than truly ruling over it, while the spread of their influence was limited to urban coastal centres.

What is perhaps rather more surprising to readers today is that we should chance upon a writer such as Victor Hugo making such a callous-sounding remark. Yet I think Hugo's comment is revealing of a broader cultural phenomenon, which we would do well to observe, which is that in the very different world of nineteenth-century France it was seen to be normal and acceptable to discuss the systematic destruction of other peoples. What we will see I believe, as we go on to look at more and more texts about extermination, is that not only was such writing seen as being acceptable, but there was also an element of necessity about the manner in which French writers talked about programmes of annihilation. It seems to me that this sense of need could have come from two places: one was a belief that one had to be honest about what colonialism really entailed – that something did lie behind the rhetorical cloak of civilisation held between the metropole and the colony, and the other is that there was a deeper ethical intimation that advocates of imperialism needed to account for the specific deaths of individuals and groups of people who would die as a result of French policies. Now these two factors may seem rather similar but it is the latter which is of especial interest to me because I think that it evokes a kind of mania which we find in writing about Algeria in which there is a certain compulsion in talking about extermination. Whereas in the twentieth century the reality of the systematic, mass killing of civilians was rarely discussed as it became a taboo in writing – if not in fact – there seems to have been a more or less converse reaction in nineteenth-century France and Algeria.

In European historical and geographical literatures, the people who would become known as Algerians were described as being amongst the most fearsome that walked the earth. British authors such as Lord Percival Barton described a people who could at times display great humanity towards others, but who contained within themselves the potential for 'savage atrocity' and an ability to 'throw aside all sense of moral obligation'.[5] For this reason, Europeans had traditionally been justified in behaving in ways which mirrored the brutal posture of Algerians towards others. The Maghreb was established as a specific moral realm in which locally appropriate forms of behaviour were sanctioned, as in Britain's raids on the coast throughout the eighteenth and nineteenth centuries which were partly aimed at the recovery of white slaves (or indeed looking much further back in time to the 'pitiless massacre of polytheists' that had accompanied the arrival of Islam in the Maghreb[6]). As the British admiral Lynam remarked on one of these sorties, 'There can be no reasonable objection ... to an occasional bombardment of a pirate town; it is a good drill for a rusty navy.'[7] In other words, violence begat violence, which was a necessary expression of European power, with there existing no distinction between combatants and civilians in this realm. Rather presciently, Barton remarks that such raids 'repressed' the 'evil' of the Barbary Coast, but that it was 'not exterminated'.[8] France would need to complete that task, in what Barton would call the Algerian 'experiment'[9]: 'that great problem in legislative science, which is to convert a barbarian race into a civilised people'.[10]

Before arriving in Algeria, therefore, French troops *knew* that their foes were implacable enemies whose warring was without limits, and it was therefore almost unworthy of comment that France's armies would have to act in a barbarous manner, adopting Lynam's policy of the general pacification of a place and its peoples, rather than specifically combating a trained army who formed but a subset of a wider population. Writing in 1884, René de Grieu looked back rather wistfully to the excitement of campaigning in the early 1830s, when French soldiers knew full well that local 'adventurers and bandits... dreamed only of massacring and pillaging'.[11] I cite de Grieu to show that French soldiers believed that they were able to access the mindset of their foes and that there was therefore an obvious and just equivalence to the manner in which the French army could also dream of the massacre.

It was certainly the case that Algerians massacred French troops in the early years of the colony. We read of quite a numbers of accounts of garrisons being attacked and their usually small number of defenders being decapitated, which other French soldiers saw as a clear message describing the way in which the Algerian dreamed of massacring them. Yet such attacks were of course very specifically directed at soldiers and at an invading army which was taking land which belonged to those who sought to defend it. This was poorly understood by many in the French army, who of course never imagined themselves in the position of Algerians (I feel sure no nineteenth-century text could be found which imagined a liberatory invasion of France by Algerians). Clauzel, however, in 1833, did understand that 'The unfriendliness of the Arabs is easy to explain, for these barbarians loathe us just as they loathed the Turks, because we are occupying a land which, is in their eyes, a place which has been theirs since ancient times.'[12] He went on to explore the possibility of whether a policy of 'extermination of the locals' was the best means to resolve this situation, but concluded that while such a policy had been enacted by the Spanish in Mexico, this had been driven by 'religious fanaticism' rather than 'political expediency'. Clauzel doubted that the French 'possessed a similar zeal for Catholicism which could drive them to treat the Arab and Berber tribes of Algeria in the same fashion'.[13]

By 1846, however, after more than fifteen years of brutal fighting in Algeria, one finds the Duc d'Aumale writing of precisely the kind of religious zeal that Clauzel could not or would not see in the French army at the inception of the colony. The Duc in fact wrote that 'the war in Africa will be a lengthy one', that it was a 'serious' enterprise' and that it was truly a form of *'Jihad* (or holy war)'.[14] It is such remarks, I suspect, which give us access to very particular exterminatory rhetorics of the mid-1840s which came at a point when some kind of 'solution' had to be found to the problem of pacifying Algeria, with the appropriation of the idea of *jihad* serving to justify a form of total war, as well as expressing a relentless determination to take whatever paths were necessary for ultimate victory in this war. Where Clauzel could not see this French *jihadism* – which linked directly to the spirit of annihilation – in the modern French soul, the Duke talks very specifically of 'the longlasting military spirit' which one finds in 'old French blood', which he hopes, 'if God wills it', 'lives on'.[15]

As well as acquiring the idea of *jihad* in their new home, French soldiers rapidly adopted the term *razzia* from Algerian Arabic. In a colonial setting where remarkably little borrowing took place from indigenous culture – I can think of no Arabic word which is more completely and frequently absorbed into French – it should not seem surprising that the French borrowed a term which described not simply a 'raid', but an attack usually of a most brutal kind in which all who stood in the path of an army were indiscriminately slaughtered. The idea which underpinned the borrowing of this term was a moral one, for what the use of the word *razzia* expressed was the notion that all that France did in Algeria – no matter how brutal – was seen as normal in terms of the local culture of politics and violence. What is more, given the opportunity, the Algerian would massacre as many Frenchmen as he possibly could.

As Alain Corbin notes, the related term 'massacre' was itself derived from an Arabic word, though there had been an extension of the term as it migrated into French for the Arabic original had referred to the killing of animals in abattoirs.[16] Looking at the etymology of the term, Corbin observes that the idea of the massacre acquired a very specific place in understandings of organised violence in France, for notions of 'torture' and 'execution' were allied to just decision-making, while the 'fusillade' had none of the 'Dionysian characteristics' of the massacre.[17] According to Corbin a culture of massacres developed in eighteenth-century French political life, but in the nineteenth century, in spite of the volatility of French politics and its civil wars, 'there was a near complete disappearance of the massacre'; or rather the classic massacre amongst the citizenry was replaced by the militarised massacre of the civilian on the streets of Paris in 1834, 1848 and 1871.[18]

The massacre was therefore institutionalised and I would suggest that it represents one of many features of nineteenth-century Francophone culture where it is hard to determine whether institutions (in this case the army) learned in the metropole and thence exported the colony, or whether the colony played a larger part in the development of metropolitan culture than we traditionally imagine. As Le Cour Grandmaison notes, 'in June 1848, certain colonial techniques were in effect imported to Paris by senior officers – such as Cavaignac, Lamoricière and Changarnier – who had long service records in Algeria'.[19] This argument is indeed extended into the twentieth century, for he goes on to contend that other Algerian innovations like administrative internment were spread through the empire and served as the basis for Vichy policy, which was an extension of what was imagined and practiced in Algeria.[20]

The razzia is a rare subject in the history of the early colony in that – in Hamdan Khodja's *Le Miroir* – we have access to an Algerian source which describes local reactions to the behaviour of the French army. It will be recalled that Khodja was familiar with European culture and initially willing to work with the French regime – as he had been with the earlier, Ottoman, empire – but his disillusionment with the practice of French rule led him to move to Paris to mount a campaign against the new colonial state. *Le Miroir* formed a part of Khodja's efforts to persuade foreign ambassadors in Paris, French politicians (especially avowed anti-colonialists) and public opinion that France should abandon the Algerian experiment, in favour of a return to Ottoman rule. Having failed in his diplomatic mission, Khodja was later to travel to Istanbul to work as an emissary of the Ottoman court.

Khodja's critique of the early colony was focused on the connections between the gap between France's espoused liberalism and the practice of her rule, especially its arbitrary and planned violence, and a more overarching ethical disgust which flowed from an observation of this hypocrisy and these disjunctures. Khodja described brutal policy choices made by commanders on the ground, which evidently went against the interests of Algerians, but held such decisions up against the aspirations of the French, noting that they 'ran contrary French principles of liberalism and therefore against the French state itself which, in principle, they symbolised'.[21] There is therefore, in Khodja, a powerful sense of the way in which it is not simply later historians who see the actualising power of a rhetorical, imagined Algeria, but also local people who suffered as the idea was promoted concurrent with brutal forms of repression.

Khodja's call to his readers in France was to acknowledge that the meaning of the early colonial enterprise was to be found more in the razzia than in theoretical texts which imagined the idea of a liberal empire. The failure to acknowledge this fact arose from an unwillingness to see that the symbolic value of the razzia lay less in 'its immediate effectiveness or its relation to the Algerian past' than in the manner in which it represented 'the dawn of a renaissance, the eruption of the future' into the Algerian present, for Khodja understood very well that a structuring of human relations took place in the early colony which would create a long-lasting dark 'colonial night'.[22]

Khodja correctly adjudged that what he was seeing in the early colony was the creation of a permanent moral world in which Algerians could be brutally subjugated and not a brief moment of militarised violence which would precede some more humane future for France's new subjects. One reason why he was able to make such observations was of course that he had known Ottoman rule well and he was able to make clear comparisons between the behaviour of old and new imperial rulers, of specific policies and a more general sense of their ethical import. Khodja acknowledged that the Ottomans had been 'despots' but he remembered that their 'iniquity' had extended only as far as the imposition of harsh taxes, unlike the 'forced exile, pillage and massacres' that accompanied French 'progress'.[23] Khodja's description of the inevitable permanency of this peculiar Algerian moral realm are made plain 40 years later in a discussion between Ideville, a prefect, and his Governor General, Chanzy, in which Ideville had wanted to dismiss a number of public officials who had committed crimes during the Commune, but was dissuaded from doing so by his superior. While Chanzy acknowledged the lack of morality inherent in not dismissing such people, he contended that the colony had no others on whom it could rely, so Ideville would need to accept the necessity of this compromise.[24]

Khodja stood as a witness to the exterminatory policies of the French in their appropriation of the idea of the razzia: to 'the theatre of horrors' which they had staged in Algiers, to the 'shameful massacre' of men, women and children by Clauzel at Blida, where breast-feeding children had been sliced apart, and the more general 'yoke of extermination and war crimes' which had been placed on the Algerian population.[25] Yet Khodja was able to see beyond such atrocities to observe that the true danger for Algerians lay in the fact that such discrete acts together constituted the formation of a broader policy environment in which 'extermination' came to be seen as a natural feature of liberal empire. In fact, Khodja noted, there were but two 'solutions' to France's Algerian problem: 'to fight to the point of either exterminating, subjugating or exiling Algerians, or the abandonment of the colony'.[26]

In stressing this element of choice and the structuring of political possibilities and their human consequences, Khodja displayed a remarkable insight into the way in which an avowedly exterminatory politics could come into being. It was not the case that politicians in France had originally planned to invade Algeria and to slaughter the native population, but they manoeuvred themselves into a decision-making process whereby extermination moved from being a theoretical possibility and one policy option amongst many, to becoming a practical and logical means of resolving a difficult situation for the French army as they reacted to changing events in the colony. There is again some tacit hint here that the idea of 'extermination' was seen as a policy which was forced upon the French by the recalcitrance of the locals, who knowingly pushed the French towards policies which would lead to their own destruction.

Unlike French colonial theorists, Khodja was alert to the possibility that ethics are as much about outcomes as they are about intentions; or, rather, that a truly ethical analysis of the French invasion of Algeria would look at the manner in which a set of avowed goals then mutated in complex ways towards a set of results which in some ways – but only in some ways – would not seem to have been implicated in the set of original decisions which organised the conquest. As Khodja put it:

> This invasion dishonoured France because it inevitably resulted in the extermination of a considerable number of beings who form a part of the human race [cette invasion est un déshonneur pour la France, puisque ses résultats doivent être l'extermination d'une grande partie des êtres qui composent la race humaine]. If the Algerian people had shared the same religion as the French, would the invaders have acted in the manner in which they did?[27]

I am most interested here in the way in which the term 'extermination' is preceded by the words 'ses résultats doivent être' for what we find in Khodja's moral imagination is a desire to think through the consequences of policies and ideas and to admit what the outcomes of such fantasies might be. As we shall see, a considerable number of French writers were also willing to make such connections.

We should also note that Khodja identified religion as the means by which Algerians could be conceived of as possessing a difference which left them outside the human community as the French conceived of it in terms of that group towards whom one should act towards as you would act towards your own. This related to Khodja's broader thesis that the injustice of the developing state of affairs in Algeria was founded upon a fundamentally misconceived and imbalanced model of knowledge. France, as he observed, mistook its power for understanding, and French newspapers believed that while they knew more and more about Algeria, that Algerians did not have access to an understanding of the French. The reverse, Khodja notes, was true, for 'the Bedouins knew of all that happened in Europe, while Europeans had no idea what went on with the Bedouins in Africa'.[28] This informed the creation of a state system where a set of ethics were conceived in ignorance (on the assumption of knowledge), whilst the consequences of such creation were a knowledge of suffering (amongst those who were assumed to be ignorant).

Khodja saw himself not as a lone or an elite voice, but as 'an echo of the facts and of my compatriots' and the chief point which he sought to make to the French was the need for them to 'listen to the pleas and the words of the inhabitants of Algeria, so that justice could be done to them'.[29] The importance of this message cannot be overstated for there is a terrible lack of sources which describe Algerian reactions to French rule, and it is critical that we note the fact that Khodja's critique of France is a moral one, based on an understanding of the way in which an ethical system was being created in Algeria, whereby ideas became policies which had results which ought to be judged ethically, rather than simply being seen as sets of outcomes which flowed from situations which were in essence dialogues with local populations. For Khodja, it was this lack of a sense of conversation which was revealing of the failure of France to adhere to the ethical principle which, more than any other, underpinned Algerian Islamic society: that of justice.

Before going on to look more specifically at French debates on 'elimination', let us consider a French response to Khodja's remarks on the morality of the razzia, from Bugeaud, architect of France's early engagement with Algerians.

On Christmas Day in 1843, the *Moniteur Algérien* included a set of three columns in which 'A Tourist' entered into a dialogue with a French army officer to try to understand how the seeming immorality of France's actions in Algeria could actually be construed as being good and just. While Bugeaud composed both halves of this conversation, it is interesting that he was well aware how the policies of the French army were viewed as being morally tendentious, and that he needed to invent an imaginary interlocutor who could only perceive the surface meaning of events until he was given access to a deeper understanding of the Algerian situation, rather than simply boldly making the case for his policies and actions himself.

The chief point of the articles was to set out a catalogue of ways of defending 'these barbarous razzias that are condemned by all the philanthropic and all the merciful minds of France'.[30] Bugeaud's first argument was a rather weak one, which was simply that Arabs did not accept 'definite boundaries', which was somewhat unsurprising since the land was theirs and it was evident that an invading army had no respect for boundaries and landholdings whatsoever. Second, Bugeaud contended that outsiders failed to appreciate the nature of war, for 'What is war in Europe and everywhere? Is it the destruction of the belligerent armies? No, it is an attack upon the interests of the people.' In other words, the razzia was simply a colonial equivalent of the seizure of 'the great towns' and 'the centres of population and commerce' found in European conflicts.[31] This would seem to be a rather inadequate defence of the razzia, since the pleas of the 'philanthropists and merciful men' had little to do with property (though perhaps they should have since the seizure of property often led directly to destitution and starvation) and everything to do with the brutality of the organised slaughter of civilian populations. Bugeaud, however, explained that such attacks were in fact a colonial equivalent of the confiscation of goods and property in European war. The essential difference of Africa, according to Bugeaud, was that people were nomadic, so some means of pacifying these 'fugitive populations' needed to be found, so that Algerians could not simply avoid battle by fleeing upon their 'camels, mules and bullocks'.[32] Tellingly, Bugeaud does not reveal to us what happened to these people when his armies worked out how to reach them, but he was determined in his assertion that it was to the razzia 'that we owe all our progress'.[33] Put more generally, such immediate violence was a form of purging that was a precondition of eventual, sustained progress, evinced in fields such as medicine.

Bugeaud's secondary argument, which initially seems quite valid, was that 'prejudices are strange things', for people wished to view the razzia as being morally different from the horrors of war in Europe, such as the bombardment of towns and the starving of the citizenry in sieges.[34] In fact, Bugeaud then dared to move on with something of a rhetorical flourish, 'The razzia is much less cruel. There is no murdering of women and children with shells, as in Europe.'[35] This was of course where Bugeaud's argument fell apart in three distinct ways: first, it was simply not true that women and children were not targeted in razzia. As a French officer remarked of one such attack, 'The carnage was frightful ... Houses, tents, streets, courtyards littered with corpses ... in the disorder, often in the shadows, the soldiers could not wait to determine age or sex. They struck everywhere, without warning.'[36] The only unusual thing about this account is the distanced, erroneous use of the pronoun 'they' for what the officer finds hard to admit is that this should read as 'we'. Second, there was surely a difference in intentionality and outcome in Algeria and Europe, since in the colony non-combatants were specifically targeted in a way that would have been seen as unusual at home (for the reasons of the dispersion of peoples and the lack of distinction between combatants and non-combatants alluded to above); and, third, Bugeaud believed that the razzia was the only effective tool available to him for the punishment of Algerians, when, as we shall see, later French soldiers and administrators would discover that it was possible to eliminate tribes and acquire large sums of money by forcing Algerians to sell their crops and livestock to pay reparations to the French. This, though, represented a later reworking of the razzia, where its violence was notionally occluded, as compared with Bugeaud's avowed aim of proudly and openly defending its violence, though it is notable in this set of dialogues that he never really discussed the effects of such violence: death could lie assumed, just as beneficent intent could be supposed.

4.6   A culture of elimination

Having tried to show how a project of elimination was construed as both practical and ethical in the early colony, I want to move on to look at the prevalence of discussions of elimination and to see how central they were viewed as being to debates about the health and future of the colony. I propose to do this by looking primarily at the French deputy Desjobert's four books on Algeria, published in 1837, 1838, 1844 and 1846, since his work contains a much more open survey of debates on elimination than is the case in the work of most of his contemporaries.[37]

Desjobert began by noting that the idea of an eliminatory colony was by no means new as it was necessarily related to earlier European imperialism.

Moral argument about elimination therefore drew upon the frames established in existing discussions. Thus Desjobert cites the question which had
been put to Desfontaines, the medical and botanical traveller who had visited Algeria and Tunisia in 1784, 'Does the character of the inhabitants of
these places lead you to think that there could be any rapprochement
between them and colonists, or do you believe that it would ultimately be
necessary to destroy them in order to occupy these lands?' to which
Desfontaines coyly replied, 'cette question est embarassante pour lui'.[38] In
other words, 50 years before the invasion of Algeria, there already existed a
sense in French culture that while elimination might be expedient, there
were compelling moral reasons for feeling that one might only allude to its
desirability, for while it did not need to be rejected as a policy option, it
could not in good conscience be plainly admitted to be a good thing. Such
a sense of there being both a knowledge of the rightness of non-maleficence
and a practical rejection of its importance as a moral category was to become
well apparent in the life of the colony.

Looking at the history of colonialism, Desjobert noted that it was commonly believed that 'the first step towards colonisation was the extermination of indigenous peoples'.[39] In the case of Algeria, if 'the complete extermination of the Algerian population' was not possible, then at the very least
the British example in America should be followed, 'with partial extermination and the complete dispersal [refoulement]' of local populations.[40] Yet, in
a work of a year earlier, Desjobert himself had queried the relevance of the
American example, for he noted that in contradistinction to America, 'all
the land' in Algeria 'was occupied', so there was no extra space into which
locals might be dispersed, which would ultimately lead to the French trying
to drive the Algerians into the desert whilst the Algerians tried to push the
French back across the Mediterranean.[41] Similar views could be found in the
work of de Gasparin, deputy for Bastia, who in 1840 wrote that 'extermination follows obviously on from colonisation', and in the *Courrier africain*,
most especially in articles by Bodichon (which Le Cour Grandmaison sees
as having tacit forms of support even as they were condemned in the
Chamber, since such discussions were not censored).[42]

By 1838, Desjobert called for honesty in admitting that France wanted to
exterminate Algerians, observing that this was a predictable outcome in
colonial situations, and, indeed, that it needed to be acknowledged that an
exterminatory 'système' had already been established in Algeria.[43] This was
partly founded on a racialised view of Algeria in which the presence of others militated against the potential success of the French: the 'Arabs would
never change their ways',[44] the Kabyles 'were still more intractable'[45] and the
Moors and Jews were incapable of working the land.[46] Desjobert admitted,
nonetheless, that the significance of these categorisations was predicated on
the French desire to take land, to exploit it and to export settlers from
the metropole. None of these things had been countenanced by Algeria's

previous imperial masters, the Ottomans, who had not therefore needed to follow the French path of 'l'extermination des indigènes'.[47]

In 1838, the governmental commission on Africa had, in theory, rejected the exterminatory path for Algeria, principally on fiscal grounds. It had noted the moral qualms associated with such a policy – with the words 'even admitting that a contemporary civilisation could consent to act in such a fashion' – but its chief point was that 'the size of forces and the huge cost' of such an enterprise would be out of all proportion with the potential benefits that might come to France.[48] Yet such a rejection was merely a repudiation of a grand, planned exercise and it by no means provides evidence that projects of extermination were not undertaken in Algeria. In fact, the language used by the commission, which lazily admitted the possibility of extermination and passed over any possible moral objections, is important in terms of coming to understand the more general establishment of a culture of elimination in the colony. As Desjobert was later to remark in 1844, although 'the question of extermination was posed in a timid fashion' in 1838, the government had already framed this discussion by asking the commission to investigate the practicality of 'the violent expulsion of the indigenous people' of Algeria.[49] While some deputies might have entertained fantasies of the peoples of Algeria willingly vacating their lands – such as Laurence's notion that 'sensing the impossibility of living alongside us, the Arab and Moor will sell their lands and move away' – the truth was that 'The colonisation of Africa leads to the Arab and French nationalities finding themselves alongside one another, which would lead to the extermination of the Arabs ['L'extermination des Arabes en était la conséquence'].[50]

In 1834, Gasparin had written that in the face of the supposed great danger of the local tribes, 'an extermination was inevitable'.[51] Renault, however, who we will remember was a great advocate of France remaining in Algeria, contested the idea that 'all the partisans of colonisation loudly argue in favour of the system of extermination'.[52] In 1839, Ardeuil's discussion of the policy options available to the French referred to what he called 'the project of extermination or expulsion of the indigènes', claiming that such an approach was 'unreasonable' and ought to be rejected.[53] He could however well understand why such discussions took place for he too was convinced of 'the ultimate incompatibility of the French and the Algerians on the basis of their differences of religion, morals, habits, language and colour'.[54]

Such discussions are also apparent in records of debates from the Chamber in 1833, where the choice between a policy of terror and extermination was put alongside more liberal options. On 8 March, Gaëtan de la Rochefoucauld announced that 'Some wish what they refer to as the razing of the soil [*balayer le sol*] of Algeria, to chase away or exterminate the inhabitants in order to replace them with Europeans.'[55] Those who advocated a 'régime libéral', which would consist of separate military and civilian authorities, were challenged by those who believed that 'it would be impossible to civilise the Arabs and that one could only hold onto their lands through a strategy of terror, and that ultimately colonisation would only be practicable when the inhabitants of the country were expelled.'[56] De la Rochefoucauld bemoaned the 'sadness of the system of terror which it had been judged necessary to establish in Algeria', which he described in some detail, noting that government ministers did not deny that whole tribes had been massacred, including women, children and the elderly, that summary executions were common and that the heads of an alleged spy and a prisoner had been impaled on stakes at one of the city gates of Algiers.[57]

Also in 1833, an anonymous English writer wrote a scathing pamphlet denouncing the French practice of imperialism in Algeria, saying that while such a plan had been fine in theory, the brutality of the French practices had robbed it of its moral virtue.[58] He wrote that 'new plans for French enrichment at the expense of Algeria are concocted every day. Without going through all of them, let us recall the projected extermination of millions of men, whose country belongs to them just as much as Paris belongs to the French.'[59] Referring generally to the French literature on Algeria, he wrote that his readers ought to be aware that:

It is said that it would not be possible for the French to succeed the Ottomans in terms of enjoying the same levels of security; that it was impossible to civilise the Arabs, and that one could only maintain peace through terror. Ultimately colonization was not practical without the evacuation of the entire population.[60]

Desjobert's analysis of the manner in which a culture of extermination developed in the colony was especially trenchant with regard to the deployment of the euphemism of 'refoulement'. As he noted in 1837,

Up until this moment in time, no one has set out in writing the means by which the Arabs are to be exterminated ['le système d'extermination des Arabes'], for wise voices have instead taken to using the term 'refoulement', without troubling the meaning of this term or looking at what it might mean.[61]

In Chapter 5, we will go on to look at the way in which the idea of 'refoulement' played an important role in stripping Algerians of their land rights – and consequently their impoverishment and elimination through famine – but in the early colony it is imperative to note Desjobert's observation that 'the word 'refoulement' served as a mask for the [idea of] extermination'.[62] We might also note that other programmes for the forced migration of Muslims from the Mediterranean basin were also being imagined in French culture at this time in areas such as the Balkans.[63]

The practical and moral similarity of the outcomes of policies of 'refoule-ment' and 'extermination' was, however inadvertently in some cases, writ large in evidence given to the Commission for Africa of 1838. As Bernard, minister of war – and Desjobert's 'exterminating angel of Africa' – remarked, 'We must resign ourselves to dispersing the indigenous populations far away; perhaps to the idea of exterminating them.'[64] The distancing language used by Bernard is rather telling here, for of course from the perspective of Paris or Algiers there was no practical difference in dispersal of locals to a notional, but literal, place far away where they would pose no further problems, and their actual disappearance from the earth. Policies of refoulement and extermination both sought the dissolution or melting away of the Algerian problem, which was the presence of local peoples in the colony.

The extent to which an exterminatory logic had begun to prevail in French thinking about Algeria is also revealed in texts which opposed such policies. Thus, we find the objections of Genty de Bussy, ancien intendant civil, to 'le système d'extermination':

> The proponents of this system say that we can gain nothing from this set of relentless, bloodthirsty fanatics, and that the wisest thing is to eliminate them [le plus sûr est de s'en défaire]. ... Today's Frenchmen would descend from the north, as did the Huns and the Vandals, in order to massacre thousands of families. Yet if we do so we shall truly be an accursed people. If there really is not enough space in this land for us and the indigènes, it would be wisest to leave them be; our pride might suffer, but our good character would be elevated. We would be admired for our actions and neither murder nor carnage would have soiled our reputation.[65]

There are a number of things which I find remarkable about these comments. First, there is the repetition of the idea that extermination was a 'système' in Algeria, implying a level of concert and organisation well beyond mere local policies and initiatives, and the evident need for a politics in opposition to elimination. Second, there is the unusual and somewhat perverted use of the euphemism 'défaire' as a means of describing planned killing. Finally, there is the question of how debates about extermination are framed in this text, for we see that the contours of thinking in the 1830s about such matters are revealed in remarks such as the claim that France's pride would suffer if it engaged in the extermination of the Algerians. The very fact that this is advanced as one of the chief objections to such policies is telling for it reveals a certain sense of desperation in the camp of the opposition, as though simpler moral arguments about the ethics of the mass killing of civilians, empire and war have been lost, so emotional claims about the character of the nation need to be called on now that other logics have failed.

Desjobert himself also offered a sense of the manner in which arguments on ethical lines were quickly discounted, writing in 1837 that he would not 'lay himself open to the ridicule of the colonists by invoking notions of equality, morality and humanity'.[66] As he noted, a profound hypocrisy lay at the heart of a culture which on the one hand prided itself on its opposition to slavery and its call for the abolition of the death penalty, while on the other it decimated and enslaved other subject peoples.[67]

By the 1840s it was quite clear to Desjobert that the evolving history of Algeria was quite unlike that of Europe, no matter how often advocates of extermination might try to cite events in Europe as moral equivalents of the massacre of Algerians, for 'European wars simply did not have the savage character which we find in the wars of extermination which we are undertaking in Africa.'[68] The scale and extent of literatures which provided moral back-up to programmes of extermination gave licence to a frenzied violence in the colony as new allusions and illusions in language served to sanction this organised fury. Desjobert remarks that,

in 1831 Marshal Clauzel had wondered if extermination would be necessary, in 1833 Marshal Soult had wondered if it were practical, and in 1846 the secrétaire de la présidence du conseil des ministres spoke of wanting to 'peacefully establish colonialism while smothering in its progress the dying waves of the Arab nation.'[69]

As Desjobert notes, 'how can one see this [last] phrase as referring to anything other than the idea of extermination?'[70] Most important of all, this continuity of aims from Clauzel and Soult through to the 1840s was indicative of the presence of a 'système', which needed to be acknowledged as one of the very foundations of France's empire in Algeria.

By 1847, doctor Warnery was spurred to write a survey of Algeria because of a pamphlet which was circulating amongst ministers that had been written by 13 Algiers businessmen on the question of 'military colonisation'. Warnery was convinced that these authors had been induced to write their text by the governor general and the army, for 'It otherwise seemed hard to understand how such serious figures and free citizens could advocate the organisation of a system, the consequences of which would have been the monopolisation of Algerian land by the army and the complete annihilation of the civilian population.'[71]

The idea of the dream of extermination reappeared in another text of 1847, Bourjolly's *Algerian Projects*. There he wrote that:

The populating of Algeria through colonialism leads imperceptibly towards a repression of the Arabs which is neither humane, nor politically sound. In reality the Arab population lies in front of our eyes, fighting for the land on which she lives from which she draws life. We cannot dream of making her disappear. The idea of extermination should be far from our minds, for we should be thinking of assimilation.[72]

While Bourjolly was opposed to the idea of extermination, he revealed its continuing appeal in the late-1840s and in his choice of language, I think, he also shows us something of its power as an idea in colon culture. The fact that he had to remind his readers that such debates referred to the bodies of living, breathing Algerians who lay 'in front of our eyes' is suggestive of how easily such bodies disappear in discussions of extermination. Such debates formed a part of what we might call a dream world of politics, which Bourjolly asked his readers to snap themselves out of.

Yet that political dream world structured the practice of politics and governance across nineteenth-century Algeria. The humanitarian impulse towards Algerians which was directed towards the idea of medicine may be criticised on many grounds, but we also need to see that it was a form of opposition to an exterminatory politics which proposed that the new colony could only truly function when the indigènes had been absented from Algeria. Across time, a residual memory of exterminatory plans remained in the manner in which the French administration was informally oriented towards measures which might lead to extermination. In debates about extermination, such as that seen in Ardeuil, both sides were agreed that the Algerians' difference and their historical status made them in some way less human than the French, which was an important step towards either formal policies of extermination or more informal mechanisms of governance which accepted demographic collapse and did not look too closely at the causes of such decline.

In such texts, Algerians were essentially construed as nature and the French as culture, which in the colony allowed for the kinds of relationships and ethics which humans have with animals. Algerians were, after all, always profoundly associated with their environments, whether they were primitives of the deserts, the mountains or the coast. In such a context, the idea of extermination made sense because policies of the killing of animals for the greater good were well known to all civilised cultures. We might recall Fanon's later description of the consequences of the confusion and suspicion which characterised the Algerian clinical encounter, in which doctors believed 'they were not practising medicine, but veterinary science'.[73] This uncertainty also had nineteenth-century antecedents in the work of the agricultural theorist Moll, who made a series of characterisations of the Algerian situation which justified exterminatory politics on Malthusian and proto-social-Darwinian terms. He spoke of a 'conflict which is but one particular expression of a common struggle between men and animals which had been taking place "since the start of the world" in which "those races which were less suited" to civilisation "would necessarily disappear as the antediluvian animals had disappeared" '.[74]

Moreover, were such policies without relevance to the future of the colony and the metropole? Were they practical expedients or might they become a means both of the colony's undoing and eventually a harsh judgement on France? Such questions were raised by Duvivier in 1841:

> For eleven years we have destroyed buildings, burned crops, felled trees, and massacred men, women and children with ever increasing fury. ... Can we believe that posterity will not judge us; that she will not castigate us in the way that Cortes and Pizarro have been judged? At least those earlier figures had the excuse of their small numbers, their religious fanaticism, and the fact that they succeeded, which excuses many things. Yet if we, who are neither small in number nor driven by religion, should fail to succeed ... then history might rightly accuse us of having massacred purely as a means of passing the time, without ever having truly known what we wanted ... and then how awfully will we be judged by posterity?[75]

Looking at French writing on Algeria from the *fin de siècle*, one finds clear continuities between discussions of elimination in the early colony and its admission as a potential policy or reality to be confronted more than sixty years after the arrival of the French (just as one finds discussions of eliminationist politics in books between the 1830s and the 1890s, such as A. Mattei's work in 1869[76]). In Pierre Cœur's 1890 polemic in favour of assimilation in the colony, for instance, one finds the author arguing that there can be room for only one religion, or culture, in the Maghreb, which necessarily leads to the claim that 'for the conqueror there are but two alternatives: to assimilate or destroy; to be assimilated or to be destroyed'.[77]

Cœur was not an advocate of annihilation, and was arguably more anti-Christian than anti-Islamic, but the space for the interpretation of what he says as being annihilatory comes in his slight confusion as to how things were going to be resolved in Algeria, how assimilation might be achieved, and the space he opens up for the potential of annihilatory outcomes. Even medicine, which was assigned the central role of directing the project of assimilation, seemed to retain a capacity to act as an agency of death as well as a motor for progressive change:

> I have no doubt ignored many of the other means by which the assimilation I desire might be achieved. No individual factor can be the sole remedy ['le rémède souverain'] or miraculous cure, but together they make up the 'régime hygiènique', in which modern medicine has finally found true therapeutic effectiveness. Whether treated by some or treated by others, some will be healed and some will die; the sick may die with treatment or he may die without; he may be healed with it, or he may be healed without it.[78]

Quite why Cœur chose to describe medicine in these ambiguous terms which linked so completely into a culture which contained within itself drives to heal and to kill, I am unsure, but I am evidently suggesting that the author was driving at a reality of the colony which was structured in Algeria from the 1830s.

This sense of history, the colony and its inhabitants, standing at a junction at which they were confronted by two possible directions in which to travel, can also be found in Henri de Sarrauton's contemporaneous *La Question algérienne*.[79] Given the subject matter of de Sarrauton's book – how the 'Algerian question' might be solved – we should not be surprised that his possible solutions were structured in a manner very familiar from the 1830s and 1840s. In fact, the core issue at stake in the Algerian question was 'this considerable indigenous population' to which France could not remain indifferent, for she needed to choose whether to treat Algerians as 'friends or foes'.[80]

In current debates, de Sarrauton could find only two possible solutions to this problem: to assimilate or to eliminate.[81] He himself wished to find some kind of middle ground between these two policy options and to do this he felt that he needed to inhabit the minds of those who stood on each side of the current policy divide, so that he might borrow from each that which would be most appropriate for the colony. He imagined the eliminationist argument thus:

> The indigènes will never accept our European mores. They will never want to become naturalised Frenchmen and if you seek to impose this, they will oppose you with all their force...They therefore need to be restrained. ...Assimilation is a chimera, for what we need to do is to disperse [refouler] these people far away and to have them systematically replaced with a French population. Algeria will never truly become French until we accept paying this price.[82]

De Sarrauton's conclusion made plain just how much his own case, from the supposed middle ground, had absorbed from this eliminationist argument. For although he expressed disgust at the barbarous – such as those who, on hearing that one Algerian has killed another in a dispute respond 'Ah well, that one less of them!' – he believed that offering Algerians the vote would have represented an equally extreme policy option.[83] He thus asserted the value of a form of enforced assimilation, noting that 'The Arab is not unintelligent. He understands very well that in this battle for existence in which we are engaged, that he will suffer a fatal blow if he does not rally behind our civilisation.'[84] This is therefore, a more or less perfect, replication of the notion that healing and killing were inextricably linked in a causal process and argument which we first saw in the 1830s. If anything, this contention is even more baldly stated by the end of the nineteenth century, with its tropicality fully intact and its now being based on a much longer history of assaults and atrocities perpetrated against Algerians to add ballast to the threat to locals if they were to take the wrong path.

Before moving on to look at the massacre at Dahra in detail, let us close here with the words of Pélissier, the man who led that assault in 1836, which relate very directly to Duvivier's thoughts on how France might eventually be judged for her actions in Algeria: 'Wherever we go in Africa, men flee and the trees disappear.'[85]

4.7   Dahra in detail

To explore further the practicalities and moral discussion surrounding French eliminationist policy in Algeria, let us return to look at Dahra in more detail, which will entail questioning some of the assumptions made in my initial narrative account. The chief reason why I want to concentrate on events at Dahra is not its infamy – for its status belies the fact that it was like so many other *razzias* perpetrated by the French – but because so many of those who took part in events there were later to write about the killings, providing us with knowledge not just of what took place, but also how such events were justified and contextualised by their perpetrators; how they saw a set of choices unfolding before them and what motivated decisions made in this process. From such sources we learn a great deal about what was seen to be reasonable to say in mid-nineteenth-century Algeria; or perhaps what it would become reasonable to say in a special, militarised, Algerian moral realm, for in reading justifications of Pélissier's actions one cannot but help think of the memoirs of Paul Aussaresses and their blunt defence of the morality of torture in the Algerian War of Independence.

For all we know about Dahra we ought also to reflect on the fact that it was only through an indiscretion that a great literature of condemnation and justification came into being. On learning of what had taken place at Dahra, Soult, the Minister of War, had wanted to keep quiet about the incident, but news of Dahra was leaked to a member of the opposition who then informed the press.

The only book written about Dahra was *Les Grottes du Dahara: récit historique*, by an anonymous soldier who described himself as 'Un ancien Capitaine de Zouaves'. The book was composed from notes taken at the time of the events at Dahra and was published in 1864 as a contribution to debates about Pélissier in the year of his death. The author was in fact a capitaine Blanc of the premier zouaves and the book's value comes partly from the fact that he had experience not only of Dahra but also other campaigns in the region which preceded and followed it.

A context in which the author tried to explain Dahra was established with accounts of earlier French engagements with other difficult tribes in the Kabyle, and the later stand-off with the Ouled Riah was explicitly compared with the army's dealings with the Sbéah in 1843. On that occasion, the French commanding officer was Cavaignac, who had tired of chasing this recalcitrant tribe in and around the hills, to the point that he was 'determined to finish things with this faithless and pitiless enemy'.[86] A wearingly familiar game of hide-and-seek culminated with the Sbéah retreating to a set of caverns, which Cavaignac proceeded to besiege. After a set of abortive negotiations, the Sbéah held a white flag outside the cave and a captain Jouvencourt took a group of men to the cave entrance. Cavaignac had warned his compatriot against such a course of action for he knew the Sbéah to be 'murderous', as was proven when the French intermediary and his men were shot at close range.[87] A justly enraged Cavaignac then ordered his men to cut branches from trees, to bundle them together and to light them, before throwing them into the caves. Blanc then went on to describe the culminating moments of this event:

> The smoke entered the caves and soon after we began to hear desperate cries from inside. A small number of Arabs fled, moving around the flames and demanding pardon. Given that the murder of the brave and generous Jouvencourt was all too recent, with the bodies of our comrades lying there before us, the colonel could have been merciless, but he resolved not to fire on these miserable specimens and ordered that the fires be put out. Those who were in the caves then exited, with the exception of fifty or so who had been asphyxiated, and the colonel offered them the peace they now requested. They offered solemn promises that they would end their brigandage, though we would later see how little their word was worth.[88]

This tale is of great importance in coming to understand Dahra for it explains – in actuality and in writing – how many of the tropes of engagement with the tribes of the Kabyle had already been established before the Ouled Riah fled into their own caverns in 1845. It was already known that these barbarous people had long relied on the caves as a means of avoiding justice, that they would use underhand tricks to try to fight off the French, and there was a casual acceptance of the tactic of burning and smoking them out of such redoubts. The death of 50 of the Sbéah was barely worth mentioning next to the horror of the end of Jouvencourt and his colleagues, and Blanc made it plain that the locals were indeed lucky that Cavaignac chose to exercise restraint in not truly punishing them for what they had done.

In a narratological sense there is something suspicious about the manner in which accounts such as this one so closely mirror events at Dahra, for an obvious implication here is both that Pélissier's decision to smoke out the Ouled Riah had just precedents and that extra levels of violence might well be necessary and deserved after the behaviour of the Sbéah (Blanc notes that the caves of the Ouled Riah were 'similar to those of the

Sbéah').[89] In this exemplary moral case, Blanc writes as though Cavaignac had actually behaved in a rather liberal and humane fashion, as though the bodies of the 50 dead locals (presumably including women and children as well as those who were notionally combatants) did not exist, as indeed they do not appear at all in this passage, in marked contrast to the descriptions of French corpses.

Blanc then went on to offer further contextualisation for events at Dahra, detailing a series of bloody incidents in the region in 1845. These included what Blanc described as 'one of those atrocities which we know one must not flaunt before the arabophiles' where two French soldiers were captured by the Kabyles and burnt alive in full view of their comrades. 'Reprisals were quick to follow', notes Blanc, 'for a few days afterwards fifteen or so Arabs fell into our hands and our exasperated men executed them on the spot.'[90] The burning of French troops was therefore incontrovertibly beyond the pale, when just months before it had been presented as a banal feature of conflict when Algerians were the victims, and we may well note the ambiguity of the term 'Arabs' deployed here, for we must take this to mean that the dead here were mainly non-combatants, for Blanc would surely have mentioned the fact if they had been soldiers.[91]

In describing the events at Dahra on 17 June 1845, Blanc stressed the choices that were open to Pélissier. Yet given the dangers to the French which were inherent in attacking the fortified caves with their narrow entrances, Blanc claimed that the only real option open to Pélissier was to lay siege to the caves. This latter course of action – which Blanc was insistent was the path which the French commander had taken – was, however, problematic, for the Ouled Riah were well stocked with food and water, and a long siege might have proven difficult in terms of 'the plans of the commander in chief'.[92] In other words, Pélissier's understanding that he did not have time to sustain a lengthy siege at Dahra, if he were to achieve the wider set of goals which Bugeaud had assigned him in this campaign, meant that he had actually needed to find a third option as a means of ending this situation.

Although somewhat labouring the point, it is critical that in Blanc's text this third option was actually seen merely as a slight variant on the peaceful siege of the caves, for in terms of the moral categorisation of what went on there, Blanc was determined to show that Pélissier's actions were driven by prudence. This third choice was in fact a threat put to the Ouled Riah that 'they would be burned if they did not accept the conditions they were being offered'.[93] Blanc implied that Pélissier menaced the tribe in this way – with accompanying visible preparations of the materials for starting a fire – simply as a negotiating tactic, yet, given the well known predilection of the French army for using fire as a weapon, this was scarcely credible. The colonel was 'exasperated' by the Kabyles who he claimed 'revelled in the impunity which their caves offered them' and who 'responded to the French exhortations with insults and exchanges of fire'.[94] This claim that the tribe felt safe in their caves was obviously not true and what is important to note here is the manner in which the cruel and ignorant foe were already beginning to be blamed for bringing their own deaths upon themselves, as though there was a moral equivalence between exasperating a French officer and burning scores of innocents alive.

It was at this point that the French troops began to burn and smoke people from the cave and, as in many accounts of these events by those who were there, Blanc used conditional language as a grammatical means of distancing himself from the immorality of an event which he notionally assures was both moral and humane [for example, 'Le nombre de fagots fut augmenté'].[95] He also assured his readers that the French troops had been ready to extinguish the flames at the first sign of submission, thus extending the idea that the tribe, and especially their leaders, were those who were at fault here. It was in fact these 'fanatical' chiefs who then chose to fire on their own people who were attempting to flee from the caves.

Once the fires went out, Blanc was quick to switch to the subject of medicalisation and care, claiming that Pélissier 'rapidly organised a system to help the hundred and fifty to two hundred survivors, most of whom were returned to life'.[96] The miraculous implications here contrast, I tend to think, with the ways in which Blanc as a writer – like all French sources on Dahra – tended to occlude the reality of the massacre as a means of evading or sublimating any sense of guilt (Pélissier was to write of the 'providential luck' that some survived in the caverns[97]). Blanc did admit that they found the bodies of men, women and children in the caves, but their deaths were neither mourned nor accorded any kind of emotion or meaning, unlike the much smaller number of survivors whose significance in the story of Dahra is writ large.

Pélissier's own account of the events – in a letter to Bugeaud of 22 June – employed many of the rhetorical strategies which we find in Blanc's later work. He too stressed his 'exasperation' with the tactics of the Ouled Riah and his evasion of direct language to describe the starting of the fires is still more striking than we find in Blanc, for he remarks 'At three o'clock, fires started all around' [A trois heures, l'incendie commença sur tous les points].[98] Throughout the letter, Pélissier was understandably determined to stress the morality of his actions, though he was often blind as to the ways in which his words might be interpreted in a much less sympathetic way than he imagined. In, for instance, describing the way in which the tribe fired on some of those who sought to escape the cave, Pélissier stresses 'the cruelty of firing on women', seemingly oblivious of the comparison that might be made with the much larger number of women who he was at that very point putting to their deaths in the caves.[99]

It is of broader significance that in his letter to Bugeaud, Pélissier felt not only a need to offer moral justification for what had happened at Dahra but also a sense of being haunted by events there. As he said, 'These were the kind of operations which one undertakes when one is forced to do so, but which one prays to God one never needs to undertake again.'[100] While Pélissier surely wrote with sincerity here, his notion that such events were somehow forced *upon him* ought to be questioned, and we ought to remember that Dahra was but one massacre in a long-standing campaign replete with such actions, in which Pélissier often had been and would be involved. Pélissier could not avoid writing of the 'horrible' sights which greeted them in the caves, but this was forcefully undermined by his assertion that the deaths of the tribe were a 'terrible lesson which they had brought upon themselves with their obstinacy'.[101]

Interestingly, an early British account of events at Dahra contradicted such accounts of the inflexibility of the Ouled Riah, for, according to Wright, far from rejecting all attempts at negotiation, the leaders of the tribe consented to leave the caves if the French 'would withdraw' but 'this condition was considered inadmissible'.[102] Given that French narratives depended upon an account of the Ouled Riah as utterly intractable for their accounts to be seen as just, it is unsurprising that this detail lies absent from their texts. Wright himself generally followed the accounts of Pélissier and his comrades slavishly, so this difference is rather telling. His book was in fact a critique of war from a Christian perspective, yet the extent to which he generally agreed with what had happened at Dahra can be gauged by his description of it as 'one of those terrible events which deeply afflict those who witness them, even when convinced of their *frightful necessity*, and when they are justified in declaring that every thing possible was done to prevent the catastrophe.'[103] My suspicion is that he was unaware how great the damage done by his own narrative to such accounts in its inclusion of the suggestion that Pélissier had chosen not to negotiate with the Ouled Riah.

The role which Bugeaud played in events at Dahra is somewhat disputed, but we are certainly fortunate to possess sets of correspondence between Bugeaud and Soult on the affair. The chief uncertainty with regard to Bugeaud is the extent to which his orders to Pélissier had specifically countenanced and imagined the use of an assault of fire and smoke in advance of the event, and precisely when such orders had become public knowledge. Derrécagaix claimed that it was only in 1850 that Pélissier revealed the existence of a memo from Bugeaud 'which covered him completely' (in other words, which specifically ordered an assault of the kind that Pélissier had mounted), whilst Commandant Grandin and others insist that Bugeaud had quickly taken responsibility for the events at Dahra, absolving Pélissier of personal responsibility for the massacre.[104]

Many contemporary accounts of Dahra certainly regarded Pélissier's defence that he was following Bugeaud's orders as being of doubtful veracity, but in a sense I find this line of argumentation a futile one, for what it was predicated on was the humanitarian hint that Dahra was somehow an aberration: that Bugeaud had behaved honourably in defending the actions of his men after the fact, though it should not be thought that he himself and French institutions in Algeria could have planned such an event. Such a view seems to me delusional for it was quite clear that events at Dahra were part of a broader, collective *système* theorised by Bugeaud and others, and there seemed little reason for believing that there was anything especially unusual about Pèlissier's actions. All that was out of the ordinary in this instance were the debates stirred by the incident in the metropole.

Bodichon wrote in 1845 that Bugeaud was, after all, a man who 'strutted around, proud in his role as the exterminator of the Arabs. He was unremitting and wholly committed to burning their harvests, killing their animals, cutting down their trees, driven to use every scourge of war'.[105] If anything, it was Pélissier who most closely followed a military code of honour in not implicating his superiors, for on 11 June 1845 Bugeaud had issued an order to Pélissier which read, 'If those rogues retreat to their caverns, you should imitate what Cavaignac did to the Sbéas and smoke them out like foxes.'[106]

In his private correspondence with Soult, Bugeaud vigorously defended both the actions of Pélissier and the principles of warfare on which they were based, determined that Soult should not be able to make a scapegoat of his colonel or to impose restrictions on the conduct of campaigning in the colony. On 25 June, Bugeaud repeated his subordinate's claim that it was the Ouled Riah who had 'forced' Pélissier towards 'such extremely rigorous action', insisting that his colonel had acted with 'all possible moderation and patience'.[107] Far from seeming to be on the defensive, Bugeaud impressed upon Soult the fact that 'The repercussions of this example would be felt terribly across the mountains and would therefore have a salutary effect', for 'within a few days all resistance would be quelled in the region of Dahra'.[108] It is important here to note Bugeaud's resort to the language of health – 'un effet salutaire' – as a means of providing ultimate moral justification for an event such as Dahra, for what he reveals is the sterilising potential of the language of medicalisation.

Under pressure from newspapers and the Chamber, however, Soult wrote back to Bugeaud demanding that he offer more information about events at Dahra, explaining that

> Public feeling about the destruction of the Ouled-Riah is running so high that if the new explanations I am asking of you do not offer me the means to calm sentiment towards colonel Pélissier, it may be that I will have to recall the officer to France and to discipline him.[109]

This was not, however, a debate which Bugeaud was going to lose; in part because of the broader principles involved with regard to control of the army in Algeria. On 14 July he forced Soult's hand in declaring that he took 'full responsibility for this act', also offering a more detailed account of what he saw as the morality of the actions taken at Dahra.[110] In doing so, he offers a series of important clarifications regarding the ethics of the decision-making processes taken at Dahra.

First, there was Bugeaud's account of why simply blockading the Ouled Riah in their hiding place and laying siege to them there was an unrealistic strategy. Given that, as we know, the tribe had taken cattle and provisions in with them, Bugeaud explained that it would have taken much longer than a fortnight to starve them from their quarters,

> and Colonel Pélissier did not have two weeks to devote to this operation for he needed to then meet up with colonel Saint-Arnaud in order to ensure the submission of the lower Dahra region. Their two columns of troops were to march alongside each other and neither of them could therefore have tarried long without harming the progress of the other division.[111]

This operational argument was promoted to the colonial public in the *Moniteur Algérien*, of July 22, which contended that 'the delay of a blockade would have endangered the success of the operation in which the columns of St Arnaud and l'Admirault were equally engaged with that of Pélissier'.[112]

This admission on the part of Bugeaud was of critical importance for it clarified the view of the French army that ethics operated in a form of layered system, in which what was generally thought to be good behaviour might easily be superseded by operational priorities which belonged to a higher category of ethics which trumped local concerns. Bugeaud railed against what he called 'the false philanthropy' of those who criticised the actions of the French army, and what he mean by 'falseness' here was really the coexistence in French culture of a different conceptualisation of ethics, in which special realms did not exist, and where certain norms were expected to apply at local and all other levels of a system. This was an ethics of immediacy – which responded to the things which it saw happening to individuals at particular moments in time – as opposed to Bugeaud's idea that ethics needed to be based upon a deeper consideration of the significance of things.

Bugeaud then filled out his conception of ethics in an interesting direction. First, he claimed that war and politics were vehicles by which a set of stated aims could be arrived at as quickly as possible. This was in fact to the advantage of both 'victors and the vanquished' for it was 'drawn out wars which ruined nations and multiplied the numbers of victims', and it thus served 'the interests of humanity'.[113] Now it should not surprise us that he called upon humanitarian ideas as a means of justifying his claims, for his ideas here did fit neatly with prevailing colonial ideas of that category of thought, but what seems unusual to contemporary audiences is that he believed that there were certain forms of warfare which were immoral and fell outside the humanitarian justification. These included poisoning, the assassination of ones counterparts and other 'perfidies', which Bugeaud contrasted with the employment of 'des moyens de force ouverte'.[114] Such views accorded with those of de Tocqueville, who also railed against the 'philanthropes' who opposed such necessities in war as the 'burning of harvests', 'the destruction of towns' and the seizing of 'unarmed men, women and children', though he too drew a line which 'excepted those things which humanity and the law of nations disallowed'.[115]

I am interested in such claims for two reasons. First, for their acknowledgement that the morality of warfare was actually more complex than Bugeaud claimed it to be, and second, because there is something of a contradiction in the logic of Bugeaud's thinking here. On the one hand Bugeaud said that any act of war could be justified through recourse to imagining the eventual outcome of the conflict – appealing to a higher abstract category of humanity rather than a duty to do justice to those with whom one engaged in the present – yet on the other hand Bugeaud demanded that only 'open' or visible forms of warfare be considered just. The calculated, hidden qualities of poisoning and assassination were rejected on precisely the grounds with which he chose to sustain his more general ethics.

Now there are local strategic reasons as to why Bugeaud was especially opposed to things such as assassination, since such methods could be associated with the brutal and primitive forms of warfare adopted by his foes in Algeria, but I also wonder if he offers us access to a broader, unseen tension or contradiction in the conduct of warfare in the colony. On the one hand, men like Pélissier and Bugeaud took great pride in the deep levels of understanding which underpinned massacres, which is what enabled them to face the corpses of those they had slaughtered, yet on another level they were not able to shake off an equal sense of unease that the moral rationale which lay behind their actions might be faulty.

Another important strand of Bugeaud's argument was his claim that the actions of Pélissier had been no more cruel than 'the bombings and famines which we subject whole towns to in European wars'.[116] Bugeaud asserted that all such things were essentially 'identical' for they represented 'war, with all its inevitable consequences'.[117] On one level, Bugeaud was evidently right, yet we have already seen that not even he subscribed to the idea that any form of action could be ethically justified in war, and the claims he made here absolve ethics of any interest in the assessment of why one choice amongst many was taken, if motives or the cloaking of one aim with another mattered, and whether distinctions between combatants and non-combatants were worth making.

In his correspondence with the minister, Bugeaud became increasingly blunt on the question as to why the massacre at Dahra had taken place, eventually admitting that the people had been killed there to set an example and to serve a wider goal. He asked Soult if he truly understood 'just how important it was for [colonial] politics and for humanity that the confidence in the safety of the caves for these and other tribes be utterly destroyed'?[118]

Set against such higher priorities, all that was exceptional about Dahra was the number of Algerians killed, and Bugeaud reminded Soult that when Cavaignac had killed 'only fifty' in 1844, 'there had been little fuss'.[119] The problem with that assault on the Sbéah had in fact been that, 'The effect produced [the asphyxiation] had been attributed to the shallow depth of the cavern and locals had not lost their sense that such places offered a means of resisting the French.'[120] Again resorting to conditional language, Bugeaud went on to say that 'what was needed was a more powerful example, related to a better known set of caves, in order to destroy a belief that lay at the very heart of all revolts in the past and which had contributed powerfully to this most recent insurrection' for 'Without the option of the perceived absolute safety of these retreats, many tribes would not have taken part in the recent revolt.'[121]

In response to Soult's fear that such an incident would serve to increase the hatred of the Ouled Riah towards the French, Bugeaud went on to explain that 'more than a hundred cases had taught him that the submission of local tribes was only true and enduring when such tribes had suffered terrible war crimes' [Bugeaud's term is actually 'maux de guerre', which would not translate exactly as strongly as war 'crimes', but which conveys an idea of sickness not present in the English].[122]

So, in many ways, if we are to believe Bugeaud's rationalisation of events at Dahra, what took place there was significantly worse than was generally believed in the French press. Where such papers concentrated on the immediate horror of the massacre in the caves, not even they would have gone as far as Bugeaud privately did in justifying the attack on the grounds that it was a premeditated, calculated and strategic strike against France's enemies in Algeria more generally. To be sure, such events were 'cruel' but 'indispensable', and they 'would in time prove to be as advantageous to those who had been defeated as to their victors'.[123] Such sentiments of course saw morality as a self-regulating/describing realm in which the question of ethics relating to dialogues between people – in this case, the dead, defeated bodies in the caves and their victorious interlocutor – did not occur to Bugeaud.

Bugeaud was to remark that 'There was nothing to add to this letter', reinforcing the impression of his operating in a hermetic moral realm driven by its own logic, and in political terms he was quite right.[124] Soult calmed the opposition, once he was sure his own position had not been endangered by the actions of Bugeaud and Pélissier, and what Blanc described as 'the hatred of a certain party, journalistic rhetoric and the affronted sensibilities of a portion of the bourgeoisie' began to wane.[125] In many ways, this in fact served to confirm the unexceptional character of events at Dahra, for this was no Dreyfus Affair, but merely a brief scandal related to a fairly typical colonial atrocity. In fact, as Behr has noted:

> The Army in Algeria quickly learned its lesson: barely two months later, Saint-Arnaud suffocated fifteen hundred Moslems in another cave, carefully left no survivors to tell the story, and in a confidential message reported to Bugeaud: 'No one went into the cave; not a soul...but myself.' Following Bugeaud's advice, the Government agreed that French newspapers should not have access to 'too precise details, evidently easy to justify, but concerning which there is no advantage in informing a European public'.[126]

Reading literatures on Dahra after the event what is in fact most striking is the backlash which spoke out in favour of Pélissier and the great number of reasons which were essayed as possibly accounting for his persecution. The very fact that there was such a plethora of explanations is suggestive, I suspect, of a certain desperate anxiety amongst his supporters to chase after what might be perceived to be a legitimate explanation for his actions: one which would fully have allayed the sense of moral uncertainty surrounding Dahra.

Derrécagaix, for instance, offered two, wholly different, possible rationalisations of the events. The first was a repetition of Bugeaud's argument that all means can be justly used in war, but the second was that it had generally not been noticed that 'the deaths of the Kabyles in the Dahra caves was essentially caused by an unforeseen accident', which was that the wind had unexpectedly accelerated the fires in the caves, panicking the tribe, who then shot wildly at each other.[127] My suspicion is that it would be harder to find an explanation of disquiet more telling than Général Derrécagaix's assignation of blame for the massacre to nature rather than man.

For Marshal Canrobert, Pélissier had made only one mistake at Dahra, which was 'to have made too much fuss about the incident'.[128] Pélissier 'had not been able to resist the temptation to adopt a literary tone in his report, with its eloquent, realist descriptions, which had described much too realistically the suffering of the Arabs'.[129] In general terms we have seen this to be untrue, for Pélissier concerned himself almost exclusively with the suffering of his own men, but Canrobert's remarks do also hint at the earlier-mentioned sense of Pélissier being haunted by the massacre, with his need to offer at least some justice to the dead with their memorialisation in writing. The literary tone which Canrobert remarked on was, after all, a mode of writing which attended to and explored the specificities of human suffering, as opposed to the prosaic qualities of the official report which baldly enumerated facts, figures and outcomes. Pélissier's openness contrasted with the suppressed descriptive urge which we find in an 1843 letter on extermination from a fellow soldier, Montagnac, who wrote 'If I let my verve for extermination carry me away, I could fill four pages for you on the subject.'[130]

Pierre Marbaud identified 'fussy and jealous politicians' and those who wished to abandon Algeria for the excoriation of Pélissier, suggesting that the colonel would be better reclaimed as a hero. He was after all a man who at Dahra had accomplished 'one of those supreme necessities on which often depend the honour and wellbeing [salut] of an army and a country'.[131] Again we see the invocation of the language of health, and Marbaud went on to repeat the literary analogy suggested by Canrobert, claiming that we should admire Pélissier as a kind of modern Hamlet who had had to wrestle with a testing moral conundrum and who had emerged a hero borne down by his own remorse at a choice he had made on behalf of the many.[132]

There were additionally a whole series of other explanations offered up for Pélissier's behaviour which drew on some aspect of life in the colony at that time, but which extended facts so far that they ought to be considered as forms of conspiracy theory. First, there was the idea, advanced by Grandin, that the press had turned against Bugeaud and Pélissier because of the support they were offering the Jesuits to establish themselves in Algeria.[133] Second, there was Berteuil's notion that the destruction at the caves had struck at the heart of an Islamist revolt which was centred on Dahra.[134] The interesting generic point about such explanations is that they resort to the idea of the hidden conspiracy in part – from our perspective – because they were unable to offer any kind of rational account of events at Dahra.

In blaming the vengeful 'dandys' of the press, Blanc too deployed a literary metaphor, speaking of the 'theatrical bombast' of such articles.[135] In an even more extreme fashion than Bugeaud he focused relentlessly on contrasting the innocence of the French and the guilt of the Algerians for what took place at Dahra, expressing a frustration with those who could not see that the 'common good' had in fact been enhanced by what had taken place there, for the 'savage men' who had 'treacherously murdered' the 'young, brave Jouvencourt as he undertook a humanitarian act', who 'had burned soldiers alive' and committed various other atrocities, had now been eliminated.[136] The contradictions in Blanc's account here were of course rife: the loading of blame for all recent atrocities onto one particular tribe, his evasion of the fact that women and children were in the caves as well as those men who had allegedly shot Jouvencourt, his rejection of any sense of proportionality and his unwillingness to countenance the idea that Algerians could themselves constitute an army exercising lethal force, in the manner in which the army to which he belonged acted. As he then remarked, 'Can the blood of the brutish Kabyles really be as precious as the blood of French soldiers?'[137] Blanc concluded that 'The terribly cruel acts of the Ouled Riah were not designed to stir generosity towards them, yet it was clear that Pélissier would never have taken such an extreme course of action if circumstances had not forced him to – circumstances ultimately forced upon us by our enemy themselves', so 20 years after the events at Dahra we see that Algerians were now blamed even more prominently for their own demise than had been the case at the time of the event.[138] Bugeaud had himself proudly argued that the massacre at Dahra was a planned assault, but as events there became historicised a more heroic narrative emerged, with much less relation to the specificities and details of that which took place in the Dahra campaign.

Blanc's memoirs are also useful in the manner in which they show the completely unexceptional nature of events at Dahra. The use of fire as a weapon of intimidation and destruction continued unabated, with Blanc boasting that he took part in the burning of 29 villages with Colonel Bourbaki in 1851. He took particular pleasure in recounting the fate of the Béni-Koufi tribe who, like the Ouled Riah, believed themselves to have 'found shelter from our fire in three large villages sited at the base of a narrow ravine', yet the villagers had not imagined that the French would be able to shell them down the ravine, burning them from their villages.[139] The cumulative effect of such assaults was that 'the Arabs, beaten back by the force of our arms, day by day began to appreciate the gifts of our paternalistic domination'.[140] Like most authors, Blanc could not resist this direct linking of brutal violence with humanitarianism. 'And if ever wild ideas of insurrection came into their minds, they had only to think of the name of the marshal who now governed over them in Algeria, M. le Maréchal Pélissier.'[141]

So far from suffering in his career after Dahra, the massacre actually provided a rapid means of ascent for Pélissier. He soon became Maréchal Pélissier, then Governor General of Algeria from 1848–54 and later the Duc de Malakoff, of whom the song 'Where's now the Mighty Malakoff?' was written during the Crimean War. Pélissier raised his own standard as Governor General bearing the message, in Arabic, 'Peace to those who submit, the sand for the unsubdued ones.' Yet in spite of Blanc's claims above about the strategic success of Dahra and subsequent campaigns of terror, there is actually remarkably little evidence that such actions diminished revolts in the Kabyle. It was in fact only a few months after events at Dahra that Bou-Maza, the clerical leader who had been the claimed ultimate target of the French, went back on the attack against the army of Africa.[142]

Given that on their arrival in Algeria, the French invaders had constantly sought to contrast their own civility with the brutal oppression of Ottoman rule in the Maghreb, it seems apt in closing this section to return to the history of the tribes of the Kabyle under that earlier empire. What even sources sympathetic to Pélissier and France admit is that groups such as the Ouled Riah had had a long history of resisting central control, of refusing to pay taxes and of retreating to caves when pursued by their imperial foes. As Derrécagaix writes, 'their Turkish or Arab masters had never dared wholly to constrain them, which had given them a great sense of confidence'.[143] It was this sense of confidence in the limits of behaviour on the part of their masters that was to be cruelly exposed by the French. In his memoirs of 1887, General Cluseret was to write of 'the système Pélissier which had been inaugurated in the caves at Dahra' and while it was in one sense true that

Pélissier's actions there systematised a method of pacification which had gone largely unreported before that time, it is important that we see that the reason why the 'système Pélissier' had such emblematic qualities was that it was so broadly representative of an eliminationist culture, before and after Dahra.[144]

4.8 Conclusion

In this final section, I want to reflect more generally on the character of a culture in which it was believed that there was a great 'necessity of reducing these people for the sake of the general tranquillity'.[145] Eliminationist cultures seem strange to us yet it is important to try to come to understand how an annihilatory politics in Algeria came to be seen not just as a difficult necessity but also as a moral good. As I have said, my feeling is that a key flaw in existing literatures is that many writers do not wish to believe that the moral universe of the Algerian colony was as strange – that is to say different from our own world – as texts from that period would seem to suggest.

We ought first to note that the Algerian example was but one in a broader colonial realm that structured modes of behaviour across the globe. Let us look for instance at a contemporaneous letter home from an officer in the English army in Afghanistan:

> We have been engaged here in the benevolent task of cutting down every tree, and burning every house we can lay our hands on, while our cattle find excellent forage among the standing crops. With a very Christian spirit we are doing all the evil we possibly can; and it will afford us the most unfeigned satisfaction if we should succeed in our zealous attempts to convert this beautiful and richly cultivated spot into a barren wilderness. That we cannot, by any wicked device, permanently destroy the vines, is a sad source of regret; but of the gardens full of mulberry, pomegranate, peach, apple, and pear trees, we hope not to leave one standing. IS NOT ALL THIS DIABOLICAL? IS THERE NOT MORE OF THE SPIRIT OF HELL THAN OF HEAVEN IN IT? AND YET WHAT IS TO BE DONE? HOW CAN IT BE AVOIDED? Our existence depends on our reducing these savages to submission, and they leave us no other means of injuring them. It better perhaps, after all, to take their trees than their lives; and all the mischief we can do to them is but a poor revenge for the barbarous massacre of thousands of our fellow soldiers, who begged nothing of them but their lives.[146]

I have included this extract because it evidently displays many of the same kinds of ideas and emotions which we find in French writing about Dahra, from the notion that a 'savage' people need to be reduced 'to submission', to the suggestion that no act of European destruction would be adequate recompense for the horrors which local peoples had unleashed upon European troops. Yet we also see here, in an especially stark form [the capitalisation is from the original], a drive to discuss in writing the 'diabolical' actions of European troops, with an acknowledgement that avowed 'benevolent' and 'Christian' motives find themselves stretched in the manner in which they were actualised as brutal and destructive deeds. Again, some higher moral code was appealed to as a means of explaining this apparent contradiction, yet it is not quite the religious realm that we might logically expect, but a primal recognition that the 'existence' of Europeans depended upon the destruction of those whom they invaded. In this case that level of devastation was restricted to 'trees' rather than 'lives', though in Chapter 5 we shall see that while the obliteration of natural resources may have seemed an oblique manner of ruining local cultures, it tended to eliminate local peoples still more effectively than overt forms of annihilation.

We should not here, though, lose sight of the fact that Dahra was very much an 'open' action in which a set of people watched another group of people die at their hands. There was arguably a symbiotic relationship between Dahras (we should use the word in the plural for we have seen there were many such massacres even if few are now remembered), deaths in famines and other 'natural disasters' founded on the destruction and confiscation of local resources, for Dahras served as exemplars in describing France's relations with Algerians. If moral justifications could be found for Dahra – and as we have seen very many such explanations were offered – then it was hard to think of forms of brutality that would not be seen as reasonable in the colony. We have seen that figures such as Bugeaud and de Tocqueville claimed that there were indeed some forms of violence which all humanity should repudiate, but this tended not to include overt, organised massacres, which were described as being somehow better than treacheries such as assassination or poisoning.

While it is true that 'l'affaire des grottes' 'became a scandal' in Paris, where it was 'denounced in the French Senate as 'the calculated, cold-blooded murder of a defenceless enemy' and in *Le Courier Français* as 'this cannibal act, this foul deed which is a blot on our military history and a stain on our flag',[147] there exists no evidence that such responses made the Algerian colony a less brutal place. In fact, as we have seen, the battle to justify Dahra on both political and moral levels was ultimately won by Pélissier's defenders, and Dahra came to symbolise the fact that killing Algerians was always a policy option. Dahra acted against the idea that 'Some kinds of killing in war are worse than others',[148] for the events were emblematic of the French army's determination not to see distinctions between civilians and combatants, nor to distinguish between the concept of targeting danger in the body of an other individual placed in front of ones eyes, and the mass killing of group conceived of as a foreign object.

Dahra also symbolised the rejection of the idea that the extremities of war entailed the making of local decisions appropriate to particular circumstances, to ensure a modicum of justice in times when the human spirit and will were stretched. Part of the tragedy of Dahra – as I too fall into literary modes – was that it was a tale foretold in which the subjects of the story were unaware of the trap into which they were falling as they beat their familiar retreat to the safety of the caves.

It is a historical mistake, I would suggest, to imagine that all genocides entail a desire to kill each and every member of a group, though that they may be a characteristic of some genocides, for a willingness to countenance the destruction of as many members of a group as is necessary for the achievement of some other aim, surely bears strong ethical similarities to other forms of genocide. In Chapter 5, we will go on to look more closely at statistical evidence of this French attack on the peoples of Algeria.

The modern theorist of massacres Thomas Nagel has written that there are essentially 'two categories of moral reasoning in war dilemmas: Utilitarianism gives primacy to a concern with what will *happen*. Absolutism gives primacy to a concern with what one is *doing*'.[149] At Dahra, as we have seen, Pélissier sought to evade any concern with absolutist reasoning, though this came back to haunt him and many others who wrote about Dahra. The question of the utilitarian character of his actions is rather more problematic. On one level, it is true that Pélissier was ultimately concerned with the broader objectives of Bugeaud's plan, and was therefore willing to do anything which moved him closer to that objective, but in theorising the massacre Nagel evidently does not mean to refer only to those who perpetrate them. There is another form of utilitarian reasoning appropriate to Dahra in which Pélissier must have made some connection between the causality of the burning twigs and the ensuing deaths of many innocents. There is rather less sense of unease about such things in the texts we have considered, so while French eliminationist action and thought is profoundly utilitarian, this is occluded through an evasion of causal realities.

There was, nonetheless, a profound difference between nineteenth-century eliminationism and most twentieth-century equivalents, for, in their writing, earlier practitioners intentionally sought to leave traces of the moral thinking which underpinned their actions. They were not proud of what they did, but they needed to show why it was necessary and they imagined that a field of ethics existed which provides some ultimate form of justification for their acts. Writing acknowledged a connection which existed between the knowledge which people had about the death of the specific native and the generalised policies of France, for it was a means of assuaging, justifying and thinking through the practical nature of the ethics of empire. It was as though a human urge existed to ensure that the ethics of subjugation were recorded, and, as we have seen, such writing could be found right across the period being studied.

The idea of extermination was in many ways a dream, but as we have already seen the problem with dreams in the imperial context is that they began to relate to reality and to the making of policy in ways which were not wholly manageable. In contradistinction to its true meaning, the *dream* seemed to have a virtue because of its sense of clarity. It proposed the idea of Algeria as a *tabula rasa* on which the French could enact whatsoever imperial fantasies they chose, making their own India in their image – and this of course involved a profound level of forgetting of the physical, human consequences of managing a project of extermination or expulsion. Rather like a contemporaneous phenomenon such as Orientalism, the eliminationist culture did not have a single theorist or an agreed upon definition, but I hope to have shown that it existed as a textual reality, and as was the case with Orientalism, such rhetorics played an important role in structuring the moral environment in which eliminationist practices could be sanctioned. Across the study of nineteenth-century Algeria there has been, I suggest, a lack of willingness to acknowledge connections between rhetorics of extermination and specific Algerian holocausts. Our concern that genocides should possess a particular structure associated with the Holocaust is profoundly ahistorical in that it both serves to hinder studies which trace the development of forms of man's inhumanity that contextualise the Holocaust, and it denies the particularity of other catastrophes. If we are ever able to make links between what people say and what they do, then the extermination of Algerians would seem to be a genocide.

I have also sought to suggest that this culture was cloaked in beneficence by its association with humanitarian health and medicine, but that this linkage was a profoundly dangerous one, for the manner in which it was structured denied all autonomy to Algerians, forcing upon them an unspoken choice in which any rejection of cultural obliteration would necessarily lead to physical destruction. We have seen that there was something of a mania for writing about medicine, and wanting to think medically, in the colony, for such literatures enabled a confrontation with insecurities surrounding death: both of the death of the European and the death of the native.

In looking at texts across the nineteenth century, I have begun to suggest how critical events like Dahra were in structuring human relations in Algeria over a broader period of time. I do not have many new specifics to add to those writers who have adjudged that the violence of the early French colony played a key role in making Algeria a permanently brutalised state. I would agree with Behr both that 'The long, bloody conquest of Algeria had established a pattern of violence that would be evoked a century later with astonishing similarity' and that 'many of Bugeaud's own theories and policies were implicitly embodied, often unconsciously, by French army officers fighting in Algeria over a hundred years later'.[150] This unconsciousness merits further exploration in studies of the embedding of eliminationist ideas within Algerian culture, most especially in the military.

Algeria, I would suggest, became something of an exemplar for the limits of colonial violence, and part of the reason why foreign literatures existed on Dahra was that there was a fascination with the means by which such massacres could be morally justified. At Guelma in 1945, across Algeria during the War of Independence, in Paris in 1958, and again throughout the recent Civil War and its continuing insurgency, it is clear that the massacre is an integral part of political life of the Maghreb.[151] Dahra helps us to understand quite why, as McDougall notes, 'Algerian history would seem to be particularly, even pathologically, violent.'[152]

Thinking backwards and forwards from Dahra, we cannot help but also think of man's earliest inhumanity in the massacres of other hominids by *homo sapiens*, and of the Armenian retreat to caves in their genocide at the hands of the Turks in the twentieth century. In 'Concerning Violence' Fanon rightly sited violence at the heart of colonialism and, like much of the Middle East, the tragedy of Algeria is that its politics have been unable to escape that tie between politics and violence, fitting Silverstein and Makdisi's thesis that in the region 'the myths and narratives that found and sustain modern national polities are situated at the intersection of competing collective memories of violence'.[153] If Parry's analysis of the state is correct – 'that violence and death are always present under the trappings of order and peace ... violence is not just present but is necessarily at the heart of governing structures' – then we need to ask why this should be the case in such an extreme fashion in Algeria.[154]

Stora was I think only partly correct when he suggested that in Algeria, 'The benefit of civilization was deployed as a favour bestowed upon the native: not physically exterminated, the latter was granted the possibility of acquiring the colonizer's superior culture.'[155] The connection he makes between the gift of civilisation and the punishment of death is astute, but I suspect that his sense of the sincerity of the offer of the gift underestimates the more powerful exterminatory drive, which could exist at times without reference to offers of benevolence.

In some sense, the desire to exterminate and to delimit the circle of care (such that it excluded Algerians from the brotherhood of man) would also seem to have originated in French frustration with Algerians. This frustration, along with forms of incomprehension and guilt, derived from both a lack of understanding of local cultures and an awareness that they would not be as easily assimilated by the French as had been presumed by the early theorists of the imperial enterprise. Eliminationism had a profound connection to that which could not be explained, for the simple might be easily purged in an ameliorative fashion, but in some circumstances, extermination may seem to be the only means of dealing with complexity. One may dialogue with the simple and be assured that they have the potential to change as you would like them to, but there is a great frustration in realising that no such malleability may exist with a complex, intractable culture, which may only be dealt with by other means. After all, it was in the caverns at Dahra themselves that the Ouled Riah had proved their own inscrutability by shooting their own people, which led very directly to the development of a sense of assuredness that the massacre had been a good thing.

5
On Attendance to Suffering and Demographic Collapse

5.1 For whom should we care?

On 11 June 1868, an unnamed French colon from the town of Ruisseau wrote to the Governor General begging for greater levels of support for local indigènes. The sentiment which underlay his letter, he asserted, was 'a sympathy for all men': 'if it would be possible to better their position, this would be a just act and an expression of well-judged charity'.[1]

It must be admitted that administrative files covering the tragic period 1868–72 do not contain many such expressions of solidarity between colons and indigènes. This was a time of terrible suffering in Algeria – as revolts, repression, epidemics and famine intermeshed – and it was also the moment Ageron called 'the victory of the colons',[2] when an increasingly united settler class finally gained supremacy over the military, and in doing so imposed still more punitive conditions upon Algerians (although we have seen that this 'victory' was restricted only to parts of the colony). What, then, drove this anonymous correspondent to write so plaintively to the administration in Algiers, on behalf not of his own class of people but with the interest of the indigènes at heart?

I wish to begin this chapter in trying to explore the possible motivations and meanings of the Ruisseau correspondent, in part as a means of pithily opening up a series of discussions which will take some time to piece together across this long chapter. In taking a statement such as this one which uses moral language, I also wish to offer a reminder that ethics will be the means by which an array of topics will become united in this chapter's concentration on the calamitous last days of the Second Empire and the inception of the Third Republic in Algeria.

My title – 'For whom should we care?' – alludes to two seminal works of ethics: James F. Childress's medical study 'who shall live when not all can live?' and Robert Goodin's essay 'What is so special about our fellow countrymen?' In this chapter, I hope to add to their discussions in offering a historical study of the ethical problems surrounding questions of rationing and those decisions which, at their starkest, allow some to live whilst condemning others to die.

In some senses there was of course a great difference between ethical discussions in the nineteenth-century colony and Childress's study of the relative merits of the allocation of 'Scarce Lifesaving Medical Resources', such as kidney dialysis, in the modern hospital. Chief among these would seem to be the fact that Childress addresses situations where the moral value of a range of choices is openly compared, whereas this chapter deals in the ethics of the construction of an implicit set of choices about the value of life and the provision of care. If famine can be construed as a disease, the brute fact is that European colons bore almost no risk of dying from malnutrition in the period 1868–72, while the lives of many Algerians were imperilled not only by famine and epidemics, but also by those actions of the state which structurally increased the likelihood of their dying from disease, poor nourishment or some combination of these two things. My suggestion will be that a utilitarian argument about the value of life lies embedded in a set of choices and non-choices made by the colonial state, and the excavation of those forms of moral action and inaction will be described across this chapter.

Perhaps the first thing to say is that in making his statement, the Ruisseau correspondent tacitly acknowledges that healthcare and welfare provision was arranged on a racialised basis in the colony. If this were not the case, there would be no need for him to make special pleas on behalf of the indigènes, for we would expect a more general plea for help for the whole of his commune. He invokes a universalist sentiment – 'a sympathy for all men' – but at that very moment reveals that such an aspiration cannot describe the structure of health provision in Algeria. This may seem a rather banal and obvious point to make, but we must remember that at that time there was still a theoretical assumption that the healthcare systems established in the colony were universal and comprehensive in the sense that, while they may have offered distinct services to different racial communities, there remained a general aspiration for the state to provide basic levels of care to all. It will be remembered that this objective much exceeded the state's hopes for the extent of its healthcare coverage in the metropole.

From the letter it is not clear whether the form of 'support' for which the writer was calling was in the form of financial aid or medical assistance. In some senses, this is immaterial because when we look at the crises of this period we realise that there was a narrow border between forms of distress which had a solely medical cause and those which derived from general impoverishment. In cases of malnutrition, infections induced or heightened through calorific deprivation, or the spread of epidemics through forced hunger migrations, it was arguable that no such borders were present at all.

This last note on hunger marches may seem to offer some clue as to why the writer should choose to compose this letter at this time. After all, we ought to be asking ourselves in what way were his circumstances sufficiently unusual for him to feel the need to plead the case of the indigènes to the supreme political authority in the colony? Was there some special horror that he saw which drove him to express sympathy for the Algerian others whom he saw around him? Was their position so terrible that it simply could not be imagined how one could not want to 'better the position' of those whom he saw lying destitute, starving and dying in front of his eyes?

In some ways, such questions lead us to return to questions of racial hierarchies under a universalist umbrella. We have already seen very many instances of organised violence and the collective diminution of health directed against indigènes, which excited little or no approbation amongst the colons, so why at this moment was the sympathy of this settler aroused? Was it because of the fact that the horror of the situation had led to its inescapability even in areas from where one was usually insulated from the realities of persecution, extermination and demographic collapse? In their state of utter desperation, had the indigènes placed their dying bodies in the parts of his town where they knew that there was plenty, in the only conceivable expression of a desire for help that they could muster? Was our correspondent describing that migration of suffering from the country to the town, whereby thousands of indigènes abandoned their famine-stricken lands to move to the towns as one last hope?

My feeling is that the former of these two possibilities is much more likely to be the case. Although the writer expresses 'a sympathy for all men', if he were like other colons at this time, it would seem much more likely that his empathy would actually only be extended so far as those indigènes whom he knew, those who belonged to his town and those who could be said to be deserving of special forms of assistance. To have spoken of the suffering Algerian indigènes *per se* would have been to have displayed a form of conceptual sympathy for those he did not see and could not know, and to acknowledge the broader existence of some kind of structural shortcomings in systems of care and assistance during this emergency (or, beyond even this, to have shown how, in large part, it was caused by the colonial authorities). This was an idea which we simply do not find in French writing from this period. Instead, I think the writer was appealing to the Governor General to find in himself some kind of conceptual sympathy for the neighbours of the correspondent, based on the trust which he needed to place in the truth of the emotional response which the writer was delivering from Ruisseau on the basis of the things which he had seen.

It does not seem plausible that the deserving poor in the mind of the writer were Algerians from outside his own area, for all the sources which we possess from this moment show that colons operated with extremely rigid pictures in their mind of their duty of care, which placed local indigènes nearest to Europeans and which almost universally rejected the claims of 'outsiders' as tenuous at best and malignant or rapacious at worst.

Attendance to suffering was therefore very much territorially determined and this spatial prism acted in tandem with prevailing notions of racial hierarchy to establish a complex of ideas about duties of care. As we saw in the first pages of this book, such ideas structured life in the colony well into the twentieth century.

Another aspect of this crisis which I think we find revealed here is a certain form of terror or fear present in the writing of those colons who did not simply hate the indigènes. Men such as this correspondent prided themselves on having a certain amount of sympathy for locals, and such enlightened and liberal views were likely to have seen them described as *arabophiles* in colon society more generally. Yet the scale of the interlocking disasters of this period was so great that liberal systems of rationing of health and welfare resources ceased to have any real meaning or effect and were utterly unable to make any kind of impact on the terrible mortality rates of Algerians. Such systems of care had been designed to offer very specific levels of support to discrete numbers of people, but when pressed, they were revealed to be utterly insufficient to offer even basic support to local populations in general.

In such circumstances it is perhaps not surprising that enlightened colonists might begin to panic for what was revealed to them in the bluntest of fashions was the fact that the systems of care they had designed offered only marginal and limited assistance to small numbers of locals, but could not now be seen to be in any way equitable or premised on any kind of universality. In fact, existing systems of healthcare had arguably simply hidden the realities of their inadequacies in that the suffering of Algerians in general was only really apparent to most Europeans if they chose to morally extrapolate from demographic and epidemiological data, but if they chose not to do that, they might imagine that the medicalised state was working effectively around them.

The moment of crisis also revealed how shallow the offer of care was even to those whom it reached, for in spite of receiving some kind of assistance, Algerians were dying in what seemed to be disproportionately large numbers, yet of course, the scale of this mortality was now revealed as being entirely proportionate with existing systems of care. It is for this reason, I believe, that the Ruisseau correspondent made his plea to the Governor General. On one level, he must have been aware that there were underlying realities of structure and rationing at play here, in which he was implicated, but if he did not make his demand to the Governor General, then he knew that he would not even have tried to remedy the faults and consequences of this system.

It may be objected that this is to read too much into this short claim, but my suggestion here also rests on the manner in which the correspondent invokes a set of moral absolutes as a means of expressing his great desire that assistance be afforded. I am of course particularly interested in the fact that the writer cites justice as being chief amongst such moral norms. Why might this have seemed like an effective ploy in such circumstances? Perhaps most obviously, such a tactic might appeal to the military's desire for peace and stability, for the suggestion of there being the perception of widespread injustice abroad might be thought to promote insurrection. Connected to this, perhaps, was the notion that those Frenchmen who knew most about Algerian society (usually soldiers rather than colons) would have been aware that a sense of justice was the guiding principle of Islamic ethics, especially those associated with health and welfare. Many in the military saw peaceful Algerians as clients who merited some kind of duty of care, though this is not to discount concurrent fashions for extermination in the army. There was also perhaps a sense of guilt apparent here arising from an awareness of the utter failure of medical universalism and the need to be seen to be making some kind of restitution to those for whom promises of care remained theoretical rather than actual. More candidly, there was probably also the reality that the Ruisseau correspondent was disgusted by the horror he saw all around him and wished for this nightmare to end.

Yet in its invocation of 'well-judged charity' there does also seem to be an acknowledgement that what is being asked for here was the strategic redistribution of a limited amount of scarce resources, and not a call for the actualisation of a universal promise. Humanitarianism was made realistic here and in such statements we begin to see signs of that shift which took place in the early Third Republic where a universalist medicalising ideology was abandoned. Drawing on Rey-Goldzeiguer, we might also note that charity at this time served as an exemplary form of naïve beneficence, for it drew Algerians into a cash economy and accelerated their economic downfall in moving them away from the collective safety of forms of barter and traditional economics.[3]

5.2   A perfect storm

The years 1868–72 saw a crisis in Algeria which was so great that it might be described as a perfect storm: the collapse of the Second Empire, the despatch of thousands of colonial and Algerian troops to fight in the Franco-Prussian War, the formation of the Third Republic with the inherent possibility that that regime might view Algeria in new ways, a whole series of rebellions across the country, growing hostility between colonial troops and the colons, whose numbers were augmented after the loss of Alsace and Lorraine, a series of crop failures, plagues of locusts, an earthquake, famines, waves of typhus, cholera and other epidemics. It is quite understandable that most historians should wish to see this period as a turning point in colonial history, for the combination of but a few of these factors would have changed Algeria in significant ways, while the congruence of this larger set of variables surely amounted to some fundamental transformation of the colony.

In this chapter, I make the case for viewing this period as a moment at which we are able to see in a particularly stark fashion many of the ways in which Algeria had been structured from the 1830s onwards, with the existence of a greater sense of continuity than might be apparent from the drama of political upheavals and changes of regime. In essence, I believe that a concentration on the history of health at this time reveals the stability of a set of premises about Algeria, many of which were as potent in their initial configuration in which they had shaped the early colony as they were in its last days. It is not my aim to offer a comprehensive political and social history of the period, but to highlight connections between such histories and that of the health of the peoples of Algeria.

I argue that this period of emergency saw the acceleration of a set of developing trends within the colony with regard to questions of race, land, capital, health, demography and extermination, though our primary concern should be to view this crisis as an expression of ideas, concepts and structures which have already been outlined in this book, chief among which were

- the power of the idea of medical imperialism;
- the limited medicalised state which ensued – and its lack of general competence or unity of purpose;
- the connection between humanitarianism and racism, in both colonial medical systems and the destruction of indigenous systems of care;
- the centrality of exterminatory designs in colonial thought and the role which medicine played in such ideas.

A significant problem for writers on this catastrophic moment has been the question as to how the various crises related to each other and what kinds of patterns of causation and structural linkage could be discerned from a time when events at local, colonial and metropolitan levels were evidently intimately connected. I cannot propose to offer a solution to this question but I will try to show how the concerns of the history of health enable us to see how these various factors came into being and how parts of their interrelatedness emerged at this time. In doing this, my argument will also impinge upon more general questions from the history of medicine which have gained greater prominence in recent years such as the relationship between war, politics, the state, modernity, medicine and health.

The thesis I will be following through the chapter is that the calamities of this time were essentially the product of welfare and healthcare systems whose notional ethical grounding was given the lie by its actualisation. In essence, there was nothing new about this moment other than perhaps an increase in scale of local suffering and the fact that it became much more apparent to men like the Ruisseau correspondent. Famine, impoverishment and the spread of epidemic disease were essentially social constructs of colonial systems of governance though, of course, their results were far from abstract as hundreds of thousands of Algerians suffered and died.

Although these events produced some anguish and sense of moral responsibility amongst French colonial and metropolitan classes, it also presented a rare set of opportunities to those groups, for at this moment of tragedy Algeria was in an unusually malleable condition, given the breakdown of local social structures which had served as the basis of resistance to French nation-building. The questions of land and the productivity of land-holding were central in this regard, for the diminution of Algerian economic productivity was both a cause and an angrily pursued goal of this ruinous period of crisis, in which there was evidently a direct connection between the life-giving possibilities of land-holding and the life-draining potential of the loss of land. Land represented autonomy and the capture of Algerian autonomy had long since been sought, leaving, as it did, indigènes wholly at the mercy of French systems of care.

Famine, as Davis, Arnold and many other writers on this period in world history observe, was in no sense a 'natural disaster', for such sustained moments of under-supply of foodstuffs were invariably the consequence of the underlying economic realities of colonies' place in a capitalist world-system. While meteorological factors played a role in the formation of famines, even such factors were invariably intimately connected to the consequences of invaders' management of local environments. Famines were, like cholera and typhus in Algeria at this time, opportunities for some, especially those who sought to remake local environments closer to their own designs.

Davis described famines at this time as the basis of a series of 'late Victorian Holocausts' and in looking in a sober way at the colony in 1872 as compared with 1868, we need to ask whether those years witnessed an Algerian genocide. My conclusion will be that this moment simply saw a continuation of an existing genocide, for the consequence of French policies of extermination, racialised governance, land appropriation, destruction of indigenous cultures and the ethical flaws in their systems of care which failed to attend to suffering, amounted to a genocidal programme. Such claims need to be made carefully but given the fact that the factors listed above amounted to sets of plans and policies which structured a state, and given our knowledge of demographic collapse in Algeria in the period 1830–72, why should this not merit being described as a genocide?

After all, as Arnold noted, although detailed census material prevents us from speaking exactly, it is clear that 'early colonial rule' often 'resulted in heavy depopulation'.[4] He cited the case of French Equatorial Africa, where a population estimated at 15 million in 1900 fell to 10 million by 1914, and was recorded at just 3 million in the census of 1921. 'Famine and related diseases were important causes of this immense human wastage. The population was worn out and worn away by forced labour, by military requisitioning and conscription.'[5] As we shall see, famine and 'related diseases' played critical roles in the Algerian demographic collapse.

We will therefore look again at exterminatory writing from this particular period in attempting to make more concrete links between texts and deeds or non-deeds for in thinking of the suffering witnessed by the Ruisseau correspondent, we must recall that a major cause of misery at this time were those decisions taken not to care for the dying, not to divert resources towards the sick and the intentional decisions made by those who insisted that this was not a disaster, a famine or in any way an exceptional set of circumstances. We will need to look carefully at the relationship between texts and realities, at theoretical writing and its impact in situations where resource depletion forced the starkest of choices upon those who controlled the levers of care. In doing this, I plan to use both previously unused archival sources and published works from this period, most of which play no part in existing literatures. This will also involve the bringing together of sets of themes which may have been described in discrete fashions in the nineteenth century, but which writers like Rey-Goldzeiguer, Gallagher and Ageron have shown need to be read as complexes. Rey-Goldzeiguer's work on this period stands as an exemplar of historical duty as she sought to piece together what she called 'the mechanisms by which the misery of the indigènes were effected',[6] a form of structural reality which she found reference to in only one anonymous administrative source from the period, and which was hampered by the wilful destruction of archival records pertaining to health by allies of MacMahon, the Governor General at this time.[7]

5.3   Rationing, sympathy and picturing the circle of care

If we look at administrative writing from this devastating period, what can we discover about the ways in which colonial officials viewed this time of terrible misfortune? Taking the reports and correspondence of prefects from across the colony at the time of the famine of 1867–68, do we find that the Ruisseau correspondent was a lone voice or representative of a broader sense of sympathy for the indigènes amongst those officials who worked most closely with them and saw their increasing levels of suffering?

What is immediately apparent from the documentation of the famine was that there was an insatiable desire on the part of the colonial authorities at local, and especially, central, levels for information. The chief concerns of the office of the Governor General – as reflected in the form of reports demanded from the localities – tended to be the numbers of people that had died in 1868, the proportion of deaths which were attributable to typhus and the other infectious diseases which accompanied the famine, the number who had died from hunger and the share of deaths between the indigenous and European populations, also divided between the numbers who were registered inhabitants of the commune and those who were outsiders. Having looked at very many such reports it is starkly apparent how rare emotional responses were to the scale of the tragedy being witnessed, or of there being any sense of responsibility to put in place measures to alleviate either the famine or associated infections. Famine was accepted as a natural disaster of a kind that tended to strike places such as Algeria and the appropriate administrative response was to record the severity of the event, so that it might be considered alongside historic famines at some later juncture. The emotional response of the Ruisseau correspondent was therefore exceptional in this regard.

The only real sign of any sense of allegiance to the dead, the dying and their families came in the manner in which there existed a sense amongst prefects of their possessing a sense of circles of loyalty. By this I mean that the nature of the administrative reports established that the prefect's inner circle of loyalty was to European settlers, followed by Jews under his jurisdiction, then Muslims, and finally, in the outer circumference of his guardianship were Muslims who had travelled to his territory, but who did not 'belong there'. In fact, 'outsiders' were often cited as the underlying cause of the medical situation of the time; a typical response being that of the mayor of Marengo who wrote to his prefect in March 1868 to allege that the typhus epidemic had been spread by 'beggars' from Cherchell and Miliana.[8]

The scale of the numbers dying was particularly striking amongst the non-resident populations, for as the prefect of the first bureau of Algiers province reported in June 1868, the already high death rates of 1867 had been greatly exceeded in 1868: 1763 inhabitants of the commune had died in June 1867, while the figure rose to 3502 in June 1868. There had been 451 deaths amongst Algerians who had come to the city from the 'territoires militaires' in June 1867 and this number rose to 5641 in June 1868, while the proportion of these two sets of mortality figures where death was attributable to starvation rose colossally, with 200 fatalities in June 1867 and 4147 in June 1868.[9] Similar results were transmitted from across the country and it is notable that in smaller towns such as La Rassauta, Miliana and Koléah, that while the numbers of local deaths stabilised or fell, they tended to continue to rise amongst incomers.[10] In the case of the latter town's report we find one of the few descriptions of such people – under the section of the *pro forma* headed 'general observations on the causes of death' which was left blank by most officials – we read that of the two outsiders to die, 'one was a child of three months who died of tiredness on the journey from Mostaganem', whilst the other 'was an unknown, found bled white on the sea-shore near to Ousadour'.

While the scale of these deaths was sometimes described as being 'extraordinary', the reports and letters of administrators offered no speculation as to why this might have been the case. It is true that these writers were somewhat constrained by both the format of the letters and reports which were expected of them, and by their relationships with their superiors, yet it is striking how little their public voices reflected on one of the great traumas of their time in the colony.

At times, they did offer additional information on the performance of doctors, the nature of the diseases faced and the behaviour of locals. In Algiers the palliative performances of various doctors were presented to the governor general's office, with the data appearing to show that in the case of the June 1868 typhus epidemic, doctors were able to cure 80–90 per cent of the afflicted, though the number of patients seen by individual doctors ranged from two to one hundred and two.[11] Such figures were said to show that 'the epidemic has not been as serious as people have suggested'.[12] Yet the epidemiological evidence did seem to suggest that the authorities faced a particularly dangerous threat. Maussy, the surgeon general at the civil hospital in Algiers reported in June 1868 that he was encountering a new strain of typhus, which he described as 'a strain *sui generis*, of a special nature, which had originally developed at the time of the army, amongst whom it spread'.[13] This admission is of especial interest for it cited the army as the chief source of the transmission of the disease (and indeed a specific 'time of the army'), as opposed to the Algerian dispossessed or pilgrims who we saw blamed in other documents for such things, and as such it represents a very rare instance of French admission that their organisation of the coming-into-being of the Algerian state may have contributed to the spread of disease; in particular that their creation of a national space from amongst the dispersed regions and peoples of North Africa, and the lack of controls of the spread of disease amongst soldiers, had introduced new potentialities for disease in the region.

In Miliana, however, Jews were identified as being central to the outbreak of typhus in the town, 'on account of the nature of their commerce and the fact that they frequent the Arab markets'.[14] There was a concerted attempt by the local authorities to move the local Jews out of the town to a set of quarantined tents 'en plein air', in part to assure others that there was 'no serious danger envisaged' to the town. Considerable negotiations took place with local Jewish leaders, who were evidently concerned at the idea that the spread of the disease could be so readily identified as emanating from their community.[15] This was indeed a foretaste of things to come, for as the century wore on, stricter and stricter forms of delineation between faith groups became apparent in the provision of healthcare.

Rey-Goldzeiguer described the stages through which such camps were established in the Tell, tracking the hunger marches into the towns, the initial attempts of the Bureaux Arabes to offer care and subsequent waves of panic amongst colons who saw these new arrivals as bearers of disease, theft and corruption.[16] Colonial authorities began to 'park' the starving in 'dépôts de mendicité', which were often managed and justified by local doctors such as Baradon in Aumale who wrote of the need to quarantine European centres of population and to ensure that these 'vagabonds' were kept at a safe distance from such centres.[17]

During the crises of 1867–68, we also hear echoes of the earlier expression of the idea of medical imperialism and the supposedly heady innocence of Algerians faced with the benevolence of superior European medicine that was so reminiscent of the texts of the 1830s. Thus, we learn that in the civil hospital of Ste Philippe, the indigènes 'showed their very great thanks for the care which had been offered to them, taking away wonderful memories of the medical establishment which had received them'.[18] Given the relative efficacy of French drugs against typhus by this point, it is surely true that many locals were grateful for such care, but it becomes harder and harder to believe that the idea of medical imperialism was so lovingly embodied in the words of indigènes, given local understandings of the broader medical consequences of the arrival of French settlers and the fact that medicine was becoming a locus of resistance.

We might also reflect on those factors which structured both the forms in which administrators were expected to describe local experiences and the manner in which they did so. While the Ruisseau correspondent was exceptional in his special pleading on behalf of local suffering, he too operated with an implicit sense that care was spatially organised in a concentric system, where the position of those who lay furthest from the colonial centre often found their placement justified on moral grounds, such as their apparent determination to spread disease. In such a system, Algerians were doubly cursed in that there were seen to be two structuring realities which underpinned a diminution of French responsibility for their care: a horizontal, temporal axis which told the story of progress identified locals as living in forms of pre-history, whilst a vertical or concentric spatial axis stationed Algerians at the outer limits of those for whom another human being should feel a responsibility to care.

It might be worth interjecting a brief methodological note here, for what I am suggesting we can see in the colony moves beyond the claim that there was a clear connection between textual description and the enactment of policy, to posit that the very structures of texts, are of critical importance to anyone who wishes to understand the realities which structured Algerian systems of care and the disavowal of duties of care. On a basic level, this is seen in the shape of the reports demanded of prefects by the Governor General – what such documents demanded of functionaries and what they asked them not to say – but I am also referring to mental structures at a very simple or epistemic level which inform our understanding of the behaviour of individuals in emergencies where resource depletion forced the starkest of choices upon those who controlled the levers of care, which included the provision of food and shelter as well as drugs.

It is important that we identify the systematic means by which care was allotted and rationed because it is crucial that the management of healthcare not be seen as being in any way accidental or overly dependent on local variables. While the picturing of care may not have been articulated on a national level as a set of policies, it was a global reality, and we ought to see that it was heavily dependent on forms of moral judgement. What I mean by this is that instinctive senses of duties of care which drew on concepts of beneficence needed to be neutralised in such circumstances – in part because rationing demanded that Europeans be provided for first – so Algerian Jews and Muslims were invariably described as maleficent presences in terms of their being bearers of disease. That a set of complex underlying social and economic realities underpinned such movements by indigènes, and that such factors were almost wholly caused by the French, was not considered, for compelling cases needed to be developed which further limited the scope of care of colonial healthcare systems.

As the crisis of the period went on, it actually presented an opportunity for the systematising of still harsher attitudes towards locals, for a malevolent chain of logic developed where locals were identified with sickness, their sickness was seen to be a natural expression of their identity, and the deaths of large numbers of Algerians was seen as something of a relief amongst the colonial classes. As historians, we might only imagine what kind of dissection of these circumstances we might have if we possessed the testimony of a Hamdan Khodja for this period. Yet the absence of such a figure need not stop us from developing an ethical critique of the manner in which famine and disease occasioned new prospects for the refinement and sharpening of existing policies of racial division and extermination in the colony.

5.4   Famine

Moving from the study of local duties of care to the global question of famine, let us begin with the response of the colonial Algerian journal the *Revue Africaine* to the calamity that befell Algeria and much of the rest of the colonised world at this time. In fact, in volumes 11, 12 and 13 of the *Revue*, covering the period from 1867 to 1869, we find no mention of one of the most important events in the nascent history of the colony. Reports on new archaeological finds are plentiful, as the detailed fabric of the narrative of Algeria's fall from historical progress in the post-Roman period was further elaborated, but French readers would have had no sense of there being the formation of historic events around writers for the *Revue* at that very moment. Such an approach was merely a part of a more general denial of the realities of famine in Algeria at this moment and, as we shall see, the Ruisseau correspondent was a great rarity, in both acknowledging the terrible circumstances of that moment and in not instinctively blaming those who suffered most for their own fates (thus providing an added opportunity to justify the denial of care).

At least 300,000 Algerians died in the famines of this period,[19] yet the blithe manner in which colonial administrators chose to ignore the realities of this catastrophe was again by no means unusual in colonial contexts. Arnold's survey of European colonialism concluded that 'famine and disease' were gifts of almost all imperial ventures.[20] As Darwin remarked – deploying the conditional language which we have seen was typical of European abnegation of responsibility for its actions – 'Wherever the European has trod, death seems to pursue the aboriginal.'[21]

I think, one might in fact be able to develop a case that events in Algeria were perceived as much more of a crisis in the metropole than in comparable cases because of Algeria's special status in the French empire. It is certainly the case that similar crises in central and west Africa received much less attention. In the famine of Niger of 1931, for instance, the colonial state was most concerned to 'maintain tax revenues for an embattled metropole, and keen to blame the "idleness", "apathy" and "fatalism" of the indigènes'.[22] Polly Hill's study of famines in northern Nigeria at the beginning of the twentieth century provides evidence of similar trends: of metropolitan refusal to offer special assistance and, in 1908, a failure to even communicate knowledge of the famine to either Lagos or London by local authorities.[23] 'When asked about the episode, the Acting Resident for Kano replied that the mortality had been 'considerable' but, he hoped, 'not so great as the natives allege' as 'we had no remedy at the time and therefore as little was said about it as possible'.[24]

In thinking about the Algerian famine, we should not forget the great determination of Bugeaud and the military government of the early colony to see the destruction of local ecologies as the single most important means of defeating resistance and proving the seriousness of France's resolve. Fire and famine were arguably the two most important weapons deployed by the French army. As General de Castellane announced in the Chambre des Pairs in 1845, the army's 'system' was to deploy its 'arsenal of axes and matches: to cut down the trees, to burn harvests, so that we would soon become masters of a populations reduced to despair and famine'.[25] Then there was General Duvivier, who remarked that 'for eleven years, we have reversed construction plans, burned harvests, destroyed trees, and massacred men, women and children with ever-increasing levels of fury'.[26]

Such remarks were not being made only in the very earliest days of the colony but well into later decades of occupation and part of their importance to discussions in this chapter is that they remind us how little we should be surprised by events in the period 1868–72. An annihilatory culture had been established in the colony and it itself acknowledged the systematic nature of its enterprise, along with its determination to use famine as a weapon in war, which it openly acknowledged bore moral equivalence with the massacre of civilian populations. In 1838, General Brossard observed that 'hunger, thirst and sickness' were the inevitable consequences of French policies of *refoulement*, and that 'catastrophes akin to those of ancient times will reappear in modern history' as a result of such policies, so from early in the life of the colony there was an awareness that famine was socially constructed and that it was caused by the French.[27]

It was Bugeaud himself who was the arch-theorist of this means of warfare as his writing throughout the period demonstrates. In February 1837, he announced to local Arabs, 'You will not labour, you will not sow seeds, you will not tend your pastures, without our permission.'[28] In 1844, it was the Kabyles who were informed that Bugeaud would 'come into your mountains, burn your villages and cut down your fruit trees', whilst in 1846 he reported that 'The strength of Abd-el-Kader lies in the resources of the tribes: so, in order to destroy his power, we must ruin the Arabs, we must burn and destroy.'[29] Such tactics, Bugeaud noted, drew on Russian strategies in the Caucasus, where they had 'attacked nature and burned the forests'.

Environmental warfare was therefore established as the potential locus of the ultimate pacification of Algeria. He who controlled nature would rule the land and embedded within such equations was evidently a form of complete identification of Algerians with the natural world, for France played the role of the most brutal and dutiful form of culture in this calculation. The identification of Algerians with nature was of course not simply a trope but also a profound reality for the logic of Bugeaud's expedient system was one which effaced in both literal and rhetorical senses any form of responsibility for the tens of thousands of Algerians who would die because of such policies. The endemic violence of France towards the indigènes was ultimately both a coordinated project which on both overt and covert levels described Algerians as subhumans whose elimination could only further the political and military goals of France and her army.

Returning to the events of 1867–68 more directly, Ageron's work in *Politiques Coloniales au Maghreb* offers an exemplary contextualisation of the famine with regard to the socio-political realities of the lives of indigènes in the colony. Of the 89,557 victims of cholera in 1867, 86,791 came from the indigènes, while the numbers of that group who died of famine and cholera across the country in the first four months of 1868 amounted to 128,812 people.[30] It was understood by authorities in Algiers and Paris that something needed to be done about this problem, but it was quite clear that the emergency relief offered – such as the 2 million francs special credit in March 1868 and the 1.1 million francs raised in voluntary subscriptions – came too late and ignored the structural realities of the situation.[31] Napoleon III wrote to the governor general, MacMahon, in April 1867, saying that 'France does not wish it that one day it might be said that she had allowed populations she ruled to die of misery', which is somewhat indicative of the metropole's central concern with the image of the problem, rather than the problem itself, yet which displayed greater levels of sympathy than the position of

MacMahon, who replied that there was no need for too great an alarm in this instance.[32]

Napoleon III may have lost office by 1871, but at least he did not need to worry that the world saw his regime as bearing responsibility for the famine of 1867–68, for evidently French discussion of the locals blame for these events had spread widely by that point in time. This is made clear in one of the first English travel guides to Algeria, Lady Herbert's *A Search after Sunshine: Algeria in 1871* in which she wrote:

I have already spoken of the famine which decimated Algeria three or four years ago, and in which such thousands of Arabs perished. People attributed it in part to the three years of drought which had succeeded each other; and in part to the way in which the Arabs had been tempted by the high price of corn to sell their usual hoards in the French markets. But from whatever cause this terrible calamity arose, its results were the same. The French clergy and sisters, with the Archbishop at their head, multiplied themselves to meet the distress; pitched tents everywhere for the distribution of provisions and clothes; and, by incredible toil and self-denial, saved the lives of many.[33]

In other words, we might say that although the famine of 1867–68 in some ways saw the death of the idea of French medicine in Algeria, in some ways it simply represented the persistence of the idea of imperial medicine as a form of moral conquest and benevolence, though in a rather slimmed-down form, for where European medicine originally asked to be judged on its capacity to create such grand ideas as universal systems of healthcare, its goodness could now be divined from simply limiting the scale of the deaths of locals, or at least being seen to be so doing. It continued therefore to be a form of delusion, but was probably more effective for now also expressing the limits to its practical goals.

The reality of the situation was that local reports like those cited above revealed the massive scale of the problem and the manner in which these events were different to previous famines because of the way in which large numbers of the starving and the dispossessed had come to the towns seeking aid, and thence to die. Ageron reports that such forced migration was particularly prevalent amongst tribes who had fought against the French earlier in the 1860s, for a consequence of their insurgency had been that their grain stores had not been as full as usual, because of the fact that they had had to concentrate on military rather than pastoral goals, and as a result of having to sell grain to pay the cost of reparations to the French, which amounted to more than three million francs in the province of Oran alone and more than six million francs across the country.[34] Tribes which had been loyal to the French stood a much greater chance of survival.[35]

Yet this did not stop French colonial sources from blaming Algerians for the famine and its results. Lavigerie, archbishop of Algiers and later Cardinal and Primate of all Africa, believed that the 'rich, extremely rich' Arab chiefs bore responsibility for this 'terrible crisis, which has spread amongst all their people because of their actions, and in which they have not sought to save their fellow Muslims'.[36] Yet, as Ageron notes, such leaders were generally impoverished at this point, either landless or deeply in debt, and in fact death rates tended to be higher in places where all tribal structures had been destroyed.[37] The French somewhat proudly claimed that the effects of the famine were in fact greater in Algeria's neighbours Morocco and Tunisia, yet this hardly stood as a testament to the grand moral claims of a medicalised imperialism.

In some later religious writing about the famine, there was a certain nervousness with regard to the fact that Lavigerie's mission of the Pères Blancs had been established in 1868 'to combat the terrible effects of famine and plague amongst the Arabs'.[38] Such uneasiness was in part politically motivated because, in the late Second Empire and evidently even more in the early Third Republic, there was an understandable reluctance to admit that the church had viewed the famine as a market opportunity, in which their own welfare services might move into those areas abandoned by the state as a means of gaining adherents and establishing a territorial foothold in the colony (not that they needed to have worried, for Gambetta was of course to remark to Lavigerie, 'anticlericalism is for France. It is not a commodity for export'[39]).

Lavigerie ignited a scandal in both metropole and colony in 1868 when he made a public metropolitan appeal for funds for the victims of the Algerian famine. This was rightly seen by the military authorities in Algiers as a criticism of their administration and the office of the Governor General was evidently incensed by the manner in which an alliance of convenience[40] formed between *colons* and missionaries when, in truth, neither group had any more interest in providing structural assistance to Algerians than did the colonial government. This conflict escalated after the Archbishop allowed the publication of a letter of 6 April 1868 in which he made the claim that Algerians were turning to cannibalism to survive the famine, with the suggestion that such horrors were the result of anti-assimilationist policies of the army, determined as it was to abandon Algerians to their fates.[41] In what Renault claimed was an *argumentum ad absurdum*, Lavigerie concluded that but two policy options now remained in Algeria: 'France must either give these people the Gospel, or rather allow it to be given to them, or drive them into the desert, far from the civilised world.'[42] Renault believed that such a claim was based upon a rhetorical strategy which posited that evangelisation was therefore the only option open to the colonial authorities, but, given the prevalence of debates on extermination and refoulement, my suspicion is that the archbishop was making a rather more direct insinuation as to the consequences for Algerians of choices envisioned in the political culture of the colony (whilst adding to that culture which saw ideas of extermination as being within the realm of policy alternatives).

We also ought to remember here that debates about religion were a complicating factor in colonial politics but they did not readily map themselves onto existing divisions. While Lavigerie relied heavily on an alliance with opposition to the army, and later to an alliance with the notionally secular Third Republic, there were of course soldiers who supported him. Following a revolt at Mostaganem in 1864, general Sonis had written home to claim that 'the Muslims will never forgive us for not being Muslims' and to bemoan the fact that in building mosques and not churches, the administration 'was reaping what it had sown'.[43] Such mosque-building reflected, I believe, that powerful welfarist tendency within the French army which rightly came to see itself as being opposed by an alliance of forces within the army and, most especially, the civil administration and the metropole.

The Society of the White Fathers went on to found two Christian villages – St Cyprien des Attafs and Ste Monique des Attafs – with survivors of the famine, who had been baptised by Lavigerie, as well as a set of mission stations.[44] Lavigerie quickly halted baptisms in the villages at the time of the famine, in response to criticism from the metropole, but the cared-for children were later baptised in Marseille and Rome.[45] Local administrators had wanted these orphans to be returned to their tribes but Lavigerie had argued that 'the children belong to me, because the life that has revived them came from me'.[46]

A Kabyle missionary at this time commented, however, that 'we talk as little as possible about religion'[47] for it was recognised that a careful path needed to be trodden by Christians. Fortunately, as Variot remarked in his work of 1887, 'conversion was not something which could be completed in a single day'.[48] As Lavigerie's words above indicate, there was no doubt that he and his *confrères* saw the events of the late 1860s as a strategic moment on which a new church might be built in the Maghreb; or, as we should more accurately say, a time when a revived Christianity could return to North Africa. Lavigerie essentially saw himself as a reincarnation of St Augustine, re-erecting the See of Carthage on his election to the Primacy of all Africa in 1884.

In his first pastoral letter on arriving in Algiers, Lavigerie had spoken of 'the flourishing North African Churches of the early centuries and of the long agonies which had followed the Vandal and Arab invasions. What was involved was not just the disappearance of a Church but the ruin of a civilization'.[49] According to François Renault, Lavigerie saw his own situation as historically comparable with the state of the bishops and monks of Gaul who had revived European Christianity after the dark moments of the post-Roman period, for 'The story of the monks of Europe would continue to inspire him, in different ways, throughout his life. He never lost the sense that it was a model for Africa to follow: for the same condition of anarchy, the same medicine should be prescribed.'[50] Renault's apt use of a medical metaphor here doubtless reflected the way in which Lavigerie saw his mission and in his story we yet again see the way in which medicine, famine, history and politics (here in its religious guise) were bound at this moment, and how medical humanitarianism was invariably a tool of historically construed brands of one of the many ideologies of imperialism to be found in Algeria.

5.5   Capital and famine

Having looked at some of the ways in which care was organised in the period 1868–72, I now wish to move on to consider three forms of structure which can be seen as both causes and consequences of Algerian suffering at this time: capitalism in the colony, land appropriation and political conflict. Such analyses will lead us towards an attempt to quantify the scale of demographic collapse at this moment and the ways in which it related to discussions of extermination.

As I have mentioned, the French refusal to consider the structural causes of famine fits Davis's account of archetypes of famine and its management in the second half of the nineteenth century. In *Late Victorian Holocausts: El Niño Famines and the Making of the Third World*, Davis recounts what he calls 'The secret history of the nineteenth century' which, as his title suggests, is that the great global famines of the period were essentially caused by the new economies and ecologies of European imperialism, and that the deaths of millions in those catastrophes amounted to a series of genocides.[51] His specific concern was the avoidable global famines and waves of disease in 1876–79, 1889–91 and 1896–1902, in which perhaps as many as 50 million people died, but does his thesis also hold for the Algerian case we are considering?[52] Davis contended that global famines were far from accidental or random occurrences at this time, but were used by Europeans as a means of expanding and deepening their empires. It was not just European governments that bore responsibilities for such disasters but also its capital, for Davis's thesis is dependent as much on the manner in which notionally decentred capitalism created a set of environmental conditions across the colonised world in the Victorian period which reduced the capacity of conquered societies to cope with environmental shocks.

While he does not consider the famine of 1867–78, he does address Ageron's work on the later famine of 1877–81, in which:

> Official attempts to minimise the famine were belied by the flood of skeletal refugees into the towns, and the governor general was forced to acknowledge the gravity of the crisis in fall 1878...But the disaster in the countryside was a windfall to the Marseille interests who controlled commerce in North African livestock products [...] the interior tribes were forced to sell their animals to livestock dealers at dirt-cheap prices.

Exports of sheep doubled while wheat and barley exports fell by half; likewise Algeria, which had exported 17,996 head of beef in the three years from 1874 to 1876, exported 143,198 head between 1877 and 1879. In order to avoid starvation, Algerians liquidated their only real wealth: their livestock.[53]

While,

Ageron has shown how the drought of 1877–81 battened upon and, in turn, accelerated the general tendency of indigenous pauperization. After the defeat of the Muqrani uprising of 1871–72, the Third Republic relentlessly extended the scope of colon capitalism through massive expropriations of communal land, enclosures of forests and pastures, persecution of transhumance, and the ratcheting up of land revenues. Indian tax extortion paled next to annual charges that sometimes confiscated more than a third of the market-value of native land. In the Kabylia, angry poets sang that 'the taxes rain upon us like repeated blows, the people have sold their fruit trees and even their clothes'. Environmental disaster simply shortened the distance to an 'Irish solution' of a fully pauperised and conquered countryside.[54]

The work of Davis and Ageron therefore leads me to think that a history of health in nineteenth-century Algeria needs to consider at greater length the scale of land appropriations which took place and the impact these had on the well-being of the population. We also need to consider what French doctors and other sources said about the value of Algerians as humans, for while we know that the population of indigènes began rising again in the late 1870s, it is clear that a demographic catastrophe occurred in Algeria in the period 1830–70. Was this a precursor of Davis's 'Late Victorian Holocausts' and how might the treatment of Algerians be compared with the lives of indigenous peoples in places such as North America and Australia, where similar demographic catastrophes took place? Was the lie of the benevolence of French medical imperialism even worse than we might imagine, in that the moral values of medicine cloaked a mid-nineteenth-century Algerian genocide?

Rey-Goldzeiguer's work certainly suggests that all of the interlocking trends observed by Ageron above were present in the earlier crisis of the late 1860s, and that that moment was critical in terms of the definitive destruction of local markets and ecologies. In their place came a 'national market' and what Rey-Goldzeiguer called the 'true end of the Algerian economy', for where previously local groups had been able to rely upon their mastery of their immediate environments, they now found themselves wholly subsumed within a greater market devoted to 'colonial exploitation' of the country's resources.[55] Local markets moved from serving their own needs to that of the metropolitan economy, with massive increases in exports of crops and livestock at that very moment when they could least be spared.[56] Rey-Goldzeiguer rightly makes the connection between the manners in which 'just as Algeria lost its economic, material basis, its human body began to disappear'.[57]

By 1870, it was possible for writers such as Auguste Dupré to claim that the woes of the Algerians were due to their fundamental incompetence in economic management and the fact that they were unwilling or unable to identify the prices that the market would bear for their goods.[58] This in turn began a chain of events that led to the local impoverishment, for they were then forced to buy goods in markets which truly were efficient, and in finding themselves unable to afford such prices, they exhausted their savings and resorted to borrowing, without making careful plans as to how this would affect their future well-being. Such a critique of Algerians was set out in a wholly castigatory manner, unwilling to see that it was precisely the French insertion of local peoples into such economic systems which led to generalised destitution, and that while the French state would lavish huge sums on the protection of the production and plans of its population, many colons delighted in watching Algerians suffer at the hands of a market, as though the state was utterly divorced from its impact.

The plurality of Algerian ecologies and markets had been systematically weakened in the decades leading up to the period of crisis, and their fragility evidently depended on the means by which a single, national space had been created by the French army and colons. We should not forget that medicine had played an important role in this story for it was the beneficent civilisational mode par excellence that had been assigned the role of dissolving the frontiers of Algerian societies and culture so that a manageable colony might be established. In terms of the history of health, French medicine failed in a double sense in this regard, for not only did it systematically undermine indigenous medical structures but it also abnegated any sense of responsibility for the medical consequences of the new national space which it itself was instrumental in creating. Medicine arguably played a role rather similar to that of colonial capitalism, for it defined itself *in toto* in only a notional or theoretical sense, as an invisible hand, denying the very distinct controllers at its tiller and their goals for the colony.

Further evidence of the interlacing of military and civil authorities in the promotion of an Algerian capitalism, which became more pronounced over time and which was able to gain greater and greater control over the Algerian populus as a factor of production, comes in Spielman's history of the Compagnie Gènevoise. Like other European corporations, the Compagnie was offered huge land concessions in 1870 (20,000 hectares around Sétif in this case) and established an agricultural business with 428 European employees and 2,917 indigènes, with annual revenues of 321,920 francs.[59] By 1923, there were only 105 European employees, with 3,571 indigènes and yearly profits of 1,277,000 francs.[60] The lessons of such increased productivity were quite clear to other capitalists.

In 1879, the French journalist Paul Bourde accompanied a parliamentary commission to Algeria and his work provides us with a rare and exemplary critique of the financial impoverishment of the country. He contrasted 'civilisational' rhetoric with the realities of punitive systems of taxation which simply did not return services to Algerians for the sums which they paid the French.[61] What is more, in addition to not providing the roads, schools and markets for which their tax revenues were notionally apportioned, the French presided over an economy which became increasingly reliant on imported French goods, whose prices rose rapidly, whilst the French progressively increased such forms of reliance and impoverishment by 'paralysing' local forms of production in failing to build supply routes and markets for these goods. In other words, Algerians were exposed to only the malign consequences of market economies as the colonial state acted against Algerians as both consumers and producers. This economic subservience was then deepened through the introduction of credit schemes and managed indebtedness of a kind which was alien to local society.[62] Rey-Goldzeiguer showed how such trends were well developed in the years before the crises of the late 1860s, with the existence of a clear divide between the economic policies of areas of civil and military control. Whilst the army sought to 'preserve social and economic structures' as a means of pacifying local populations, the civil authorities acted in precisely the manner which we would associate with the forebears of the 'victory of the colons', raising taxes and encouraging dependency at precisely those moments when Algerians were faced with the greatest challenges in terms of drought and disease.[63]

The capitalist route to the deepening of French control in Algeria was particularly appreciated by Saint-Simonians such as doctor Warnier and Devigne. What I think Walter Benjamin taught us about the followers of Saint-Simon was that it was fallacious to really think of them as men of the left. They were deeply attracted to the progressive, expansionist goals of an optimistic socialism, but it was growth and progress *per se* which were their chief motivations, rather than an attraction to the provision of universal justice. As Benjamin noted this was nowhere more apparent than in the establishment of railway networks, for which Saint-Simonians had an evangelical zeal, in spite of the social and political questions which were raised in the establishment of such monopolies and oligopolies. This was apparent in Devigne and Warnier's work advocating the creation of a railway network across Algeria. Their first proposal was that such a scheme should be a public-private partnership, in which the state granted a concession to run a railway to a private company, as well as a gift of 10,000 hectares of land running alongside the lines.[64]

Such a railway represented 'the long-searched-for solution to the active and prosperous colonisation of Algeria', for 'at the moment at which our armed conquest can stop, we must open up a peaceful form of conquest through the greatest instrument of progress known to man'.[65] The mania of such schemes is suggestive of how far ideas about Algeria had outstripped the more realistic appreciation of things found in the writing of men such as Joly. Where, in 1833, he had seen that Algerians evidently perceived the French to be predatory conquerors, Devigne and Warnier worked on the blind assumption that after a further 21 years of fighting, that Algerians would be able to intuit the progressive qualities of the French, even as 10,000 acres of land was expropriated for the gain of a private company. Devigne and Warnier reasoned that the Muslims of Algeria 'would open their arms to this great work of regeneration that we are undertaking, in part because of the staunchness of our defence of their brothers of the Orient in their battles with Russia'.[66]

Devigne and Warnier's delusion that the Crimean War might lead Algerians to come to see France as an international defender of Islam was matched by another conceit of theirs whose origins lay in that War. Given the current closeness of France and England, they wondered aloud whether it might not in fact be plausible that English capitalists, 'less timid than those of France', might be attracted to the idea of financing an Algerian railway.[67] They reasoned that Algeria could become a part of a new route to India and a gateway from Europe to 'the riches of south Asia'. In fact, Devigne and Warnier's invocation of this potential threat or form of competition was unnecessary, for in the 30 years after their book, the French state established a railway system across the country, in precisely the same manner that an oligopolistic railway network had been established in France in the 1840s and 1850s, with a number of those French companies taking control of Algerian lines.

The point about such development is not just that it was driven by the Saint-Simonians' fantastic dreams of the potential of progressive imperialism, but that it demonstrated the contiguity of capitalism and the imperial state (as well as showing the manner in which areas outside medicine could be used as emblems of the progressive idea). The Algerian colonial state was militarised but it also served the interests of capital, and in the first four decades of the colony such interests tended to be neatly balanced. At critical points in the history of the colony, and discussions of its future, the interests of French capital were well represented, as we know from pamphlets like the 1834 *Petition of the Colons of Algeria, Followed by that of the Traders of Marseille and the Deliberations of the Municipal Council and Chamber of Commerce of that Town* which fed into discussions of the 1834 royal commission on the future of the colony. The merchants of Marseille were enthusiastic advocates of the idea of Algeria as a regional empire, just as the silk traders and manufacturers of Lyon were later to conceive of Laos in a similar fashion.

The broader effect of the development of the railway was intimately connected to the envisioning of Algeria as a land of new productive possibilities, for, as Cherry observes:

> The ecology and appearance of the region were drastically altered by the introduction of European plants, crops, animals and diseases as well as

by the new transport infrastructures. Existing species were seriously depleted and soil structures were fundamentally changed by European systems of farming and cropping.[68]

The remaking of the Algerian landscape was never, therefore, an innocent enterprise, even leaving aside the morality of land expropriation. Walter Benjamin's famous critique of the Haussmannisation of Paris has its twin in the public health infrastructure of Algeria from that same time. Benjamin had suggested that beneath the rhetoric of public benevolence and humanitarianism, Haussmann's remarking of Paris along more 'sanitary' lines – with the establishment of new systems for water and sewage, the building of parks, slum clearances and the widening of boulevards – had had as its chief motive the expulsion and dispersal of the concentrated danger of the urban poor to the outskirts of the city, and the creation of a new street plan which would stop urban revolt by preventing the erection of barricades and facilitating the ease of movement of troops through the city. Paris would therefore become a bourgeois city and a celebration of the culture of that class. We see exact equivalents of such motives and consequences in the establishment of public health measures in Algeria, such as the vaunted system of canals which were built with the stated aim of bettering communications across the country and preventing the ease with which infectious diseases such as malaria spread.[69] What such canals also functioned as were lines of defence against the indigènes and an important means of pacifying a dangerous landscape for the French, for just as the space of the French capital was re-imagined in the 1850s and 1860s, the project of public health was engaged in a similar scheme for the national space of Algeria.

As Porch noted, an analogous case, not acknowledged by the French, came in Saharan settlements in eastern Algeria, which were regarded by the French as extremely healthy places as colonial postings, until they 'replaced the primitive wells and foggars – lateral shafts dug into the sides of limestone hills – of the Algerian oases with artesian wells; these wells flooded the oases, creating lakes in which mosquitoes could breed.' Thus, the endemic malaria of towns like Murzuk was allowed to spread across the eastern Sahara.[70]

The priorities in French governance and production were such that the consequences of such plans on the lives of Algerians were intentionally ignored. Like the broader notion of empire and the idea of medical imperialism in Algeria, this was a system of ethics which drew meaning only from its intentions and never from its consequences. As Polanyi says of imperialism:

> The catastrophe of the native community is a direct result of the rapid and violent disruption of the basic institutions of the victim (whether force is used in the process or not does not seem altogether relevant). These institutions are disrupted by the very fact that a market economy is foisted upon an entirely different organised community; labor and land

are made into commodities, which again is only a short formula for the liquidation of every and any cultural institution in an organic society.[71]

This can be seen in a more directly medical fashion in the 1870 doctoral thesis of Paul Bonnet on the cholera epidemics of Algeria in 1865, 1866 and 1867. Given that a global cholera pandemic was ongoing it seems somewhat surprising that Bonnet advocated the ending of all *cordons sanitaires* and quarantines in Algeria, on the basis that it had been proved that transmission of cholera did not take place directly from person to person. This established the inefficacy of such preventive measures, which Bonnet described as being 'barbaric' because of the manner in which they 'hobbled commerce and interchange between countries'.[72] While Bonnet's reasoning as to how cholera was most likely transmitted was correct, he evidently failed to perceive how useful such measures could be as part of a broader public health strategy aimed at diminishing the spread of cholera, for an assessment of the risks involved led him to conclude that the facilitation of commerce trumped the need for collective security.

In such a view, we can see how the interests of the army and of capital might diverge at times. Grove's work on *Green Imperialism* provides, I think, a means of understanding a set of environmentalist traditions which still had some resonance amongst the French army in the mid- to late-nineteenth century. Drawing on examples from French imperialism in the Pacific, Grove claims that the conservationist instincts of earlier colonial authorities were much greater than one might expect, for there was perceived to be an implicit tension between, on the one hand, capital and its tendency to disrupt local societies and markets, and, on the other, environmental management, which served to assure local peace and stability.[73] Yet while the conflicting drives of capital and the military might be easily resolved in favour of the governmental power of the army in small, isolated colonies thousands of miles away from France, we have already seen that the army faced a great battle with competing forces in the Algerian colony, so it should not seem at all surprising that at that moment of military weakness in the closing days of the Second Empire, the forces of capital struck decisively to force a new equilibrium of power and the management of the Algerian environment.

It is also true that doctors could rail against the marketisation of Algerian society. Writing in 1851, Bodichon alleged that 'the great landowners were the enemies of Algeria, for the manner in which they left their lands uncultivated and brought about the depopulation of the country'.[74] I am uncertain here as to whether Bodichon uses the 'dépeuplement' to refer to the forced migration of Algerians to cities as a result of land confiscation and uncultivation or whether he is referring to a more sinister notion of planned demographic decline. I believe the former interpretation is correct, but the latter interpretation needs to be borne in mind.

Looking at the lives of medical professionals practising in the 1880s and 1890s, we shall also see that doctors objected to the manner in which markets for medical services developed in Algeria. Such issues were of course of great concern to doctors in the metropole, but at precisely the moment that centralised, state-regulated bodies managed to secure a monopoly on medical provision there, doctors in Algeria found themselves increasingly competing with medical entrepreneurs who had come to Algeria to try to build up a client base, with mission doctors, and with so-called 'médecins de colonisation' who were paid by the state in a revival of the idea of medical imperialism. As the grip of the state tightened in France, it loosened in North Africa.

5.6   Land appropriation and famine

In Ageron's 'victory of the colons', the year 1871 marked a turning point in the history of the colony, for after the failed revolt of the indigènes in that year, the colons were determined to assert their own group identity and influence upon decision-making and to exact a terrible revenge upon locals. Huge swathes of land were expropriated – particularly for new colons from Alsace and Lorraine – and colossal reparations were demanded, which amounted to up to 70 per cent of local capital.[75] Local newspapers were as blunt as had been the writers of the 1830s: *L'Algérie Française* of 21 May 1871 demanded that 'the insurgents be deprived of their lands, their animals and their goods of all sorts ... what we demand is that they be pushed back and confined'; while *La Seybouse* of 17 June announced that 'Terror must fall upon these lairs of killers and terrorists. The repression which we enact must be so great as to engender legends of its brutality which will ensure the future security of our immigrants.'[76] 'The Arab needed to submit to the will of the victor; to assimilate into French civilisation or to disappear', for this was 'the law of colonisation'.[77]

The scale of this confiscation of land and goods was undoubtedly novel for while Bugeaud had often advocated the destruction of local crops and lands, Ideville (a prefect of Algiers) observed in 1884, that Bugeaud had believed that only limited quantities of cattle and corn might be obtained as reparations from the indigènes.[78] This approach could be contrasted with that of the Comte de Gueydon, Governor General in 1871, who did not believe his council when they reported that it would not be possible to obtain money from the Kabyles as reparations for the war they had just lost against the French, for such people 'had nothing'.[79] The gambit which de Gueydon posed to the Kabyles was to either accept his demands or to 'have their fields ravaged', so – in a neat inversion of the Treaty of Frankfurt – France was able to express her continuing power in the colony just as she was emasculated through defeat and the payment of reparations to Germans at home.

Most fertile or otherwise productive land was appropriated from Algerians in the nineteenth century. In the early decades of imperialism land was bought, sometimes on a voluntary basis and sometimes in an enforced fashion which was said to take security factors into account, while as the century wore on, land came more and more to be seen as a legitimate spoil of war, confiscated in battle or in reparations arrangements. A select number of tribes, such as the al-Qasimi of the *zawiya* of Hamil, were able to hold on to their lands in a transaction whereby sovereignty was traded for guarantees of security and stability for the French, but we have also seen that those tribes who most actively challenged the French were liable to find themselves landless.[80] A more typical example was the Hachem tribe in the département of Constantine, who were dispossessed of 50,000 hectares following the insurrection of 1871. As Spielman notes, 'in return they were offered 20,000 hectares of sterile land which could only be worked with hydraulic technologies which were never put in place.'[81] Writing in 1870, Dupré observed that the sénatus-consulte, which had been assigned the task of analysing the crisis in Algeria, 'far from halting local dissipation, accelerated such trends, for famine gained a hold and the landless Arabs began to die *en masse*'.[82]

The punishment of such groups was criticised by Mélia, a former chef de cabinet of governor general Luteaud, in the celebrations of the centenary of the conquest in 1930:

> They live outside of a world of freedom (in terms of the length of their military service and the barriers which are put in place to their advancement). They are excluded from the purchase of all colonial land ['terres de colonisation']. They live outside the world of our system of justice, punished by repressive tribunals and criminal courts.[83]

Therefore, important amongst these forms of systematic repression which had been established by the time of the Third Republic was not only the confiscation of Algerian land from Algerians but also a prohibition on Algerians re-purchasing this land. We might ask what kinds of repression were at work here, for it seems to me that on a very basic psychoanalytic level that here the French also seemed to be repressing the memory of their confiscation of the land and their knowledge of its true owners. In *The Intimate Enemy: Loss and Recovery of Self under Colonialism*, Ashis Nandy used the example of the British in India to show that on a psychological level (if not on the level of political and social repression) the effects of empire were as great on the metropole as they were on the colony. He argued that one of the central features of European imperialism was an unshakable sense of guilt which came from the knowledge that others' lands had been stolen, in a manner in which the self would resist if it were her lands confiscated, and in a subconscious sense Europeans were well aware of the hypocrisy of a situation whereby they celebrated their societies as centres of moral progress yet oppressed their colonial subjects. An understandable response to this, according to Nandy, was not the forgetting of the crime of land confiscation, but an endless celebration of empire in public rituals and other events which drew the wider nation into a sense of collusion in the imperial enterprise.

This was precisely what took place in France in the great festivities of 1930 – which included the issuing of commemorative stamps (spreading the message around the world as well as France), great exhibitions, publications and posters – and Mélia is quite correct in identifying the land situation as one of the chief injustices which was elided at that time. It would have been one thing to have taken land and then allowed it to pass into a market for sale, but the exclusion of Algerians from the purchasing of their own land was precisely the kind of phenomenon that interested Nandy. This stricture implied that there must be some cogent moral or political logic which underlay the decision to prevent such sales, for in normal circumstances it would of course be wrong to stifle the market in this way. It was just such twisted forms of logic which interested Nandy, for in his schema the prevention of such sales had its origin in the guilt which the French assumed for their theft of Algerian land, and the skimpy logic of the moral reasoning which had claimed to underpin such acts. We might also add that such forms of warped thinking were especially powerful with regard to Algeria for its special place in the French empire was of course predicated on the idea that unlike other colonies, it was both separate from France and a part of the nation, a confused state readily apparent in texts which reflected French consciousness of Algeria.

A question we might then ask ourselves is what kind of moral cases the French developed for the appropriation of Algerian land in the nineteenth century. A particularly revealing set of answers was provided by the Saint-Simonian doctor Auguste Warnier in 1863. In answer to the question, 'Does the land belong to the indigènes, and can we, without accusation of plundering, claim a part of it for ourselves?' Warnier argued that the French must remember that:

> There are two populations in Algeria: the Berbers, who have been there since the time of the Romans, and the Arabs, who arrived as invaders driven by the Koran and the sword in the seventh century, so the land is in fact governed by two different systems of rule, one Berber and one Arab.[84]

One can, I think, see the direction which Warnier's argument is travelling here in his use of both race and history to try to establish a set of 'cultural' arguments which could be attached to the broader mission of moral imperialism and serve to justify the rather baser goal of stealing land. He went on to admit that

> Berber property rights have their origin in Roman law, like that of France, and wherever land is occupied by Berbers it is held in private ownership as is the case in France. It should thus be the case that the colonial project

should deduct the 3.5 million hectares of land owned by Berbers from its total and not lay claim to the smallest sliver of these territories.[85]

Therefore, the model which was invented by the French equated a right to ownership of land with an adherence to the Roman model of law, a system of private property rights organised under such law and the kind of settled, sedentary lifestyle associated with such cultures.

As we might imagine, Algerian Arabs did not fare so well in this schema for the Arab invaders of the seventh century were accused by the French of stealing fifty-six and half million hectares of land from 'the poor Kabyles who they had reduced to a miserly 63 acres of land per head of population'.[86] It thus followed that 'If the French government of Algeria had succeeded that of the deys of Algiers, what no one would contest, was that it was morally and legally entitled to take whatsoever land it wanted from the Arab share for the purposes of European colonisation.'[87] While we might admit that this argument had a certain internal logic in its deployment of history, it is shot through with flaws. First, Berber rule preceded that of the Romans in North Africa; second, it was simply untrue to characterise 1100 years of Berber/Arab relations as hostile; third, the categories of Berber and Arab were nowhere near as stable as Warnier presumed; and, finally, the story of the mass expropriation of Berber lands by Arabs was untrue. There was of course a much more pragmatic reason as to why the French were apt to conceive of such a scheme whereby Berbers were said to have rights to their lands, whilst Arabs were not; or we might say, that there were a mere three million reasons as to why Berbers could keep their mountainous, unproductive territories, and fifty-six million reasons as to why the French desired Arab lands. We ought also to note that French characterisations of the history of law here were also erroneous, for it was nonsensical to isolate complex systems of property rights as pertaining only to Roman law, for such rights also existed in Islamic law, yet the problem was that the French were so desperate to impose racial dichotomies on 'Berbers' and 'Arabs' that while one group were good, settled, peaceful and lawful, the other needed to be bad, nomadic, warlike and lawless.

Views such as those of Warnier had been apparent from the inception of the colony. In the critical parliamentary debate on Algeria in March 1833, Marshal Soult, the Minister of War, expressed the view 'that it is not as simple as one might think to civilise peoples who operate with different religious principles to our own, and who regard themselves as the owners of their land'.[88] Soult's remarks were made in the context of discussions in which commercial interests and the potential for the capitalist exploitation of Algeria were central to the idea of expropriating land from these other people who 'regarded themselves' as its owners. As Joly said in that same debate, 'It is important for France, for the chamber and, in particular for the commercial interests of the Midi, that we know what will become of Algeria.'[89]

Joly went on to note that on the question of the appropriation of land, local people 'had seen in this requisition an extraordinary departure from generally agreed laws; they do not see us as friends, as brothers and bearers of civilisation, but as conquerors who wish to plunder their riches through summary expropriation'.[90] The combination of Soult's and Joly's remarks appeals to me for what it reveals is the intertwining of the interest of French capital and the army in the Algerian adventure. Those two forces might have many causes to clash in the system of 'mixte' administration, but at base, many of their interests had been the same. We might also note that it was an elision of precisely the realities which were recognised by Joly that Warnier was attempting in his later work, as were colonists in general by that time.

The broader significance, of course, of such discussions to a history of health, was that the question of land and its ownership was of critical importance to the welfare of the peoples of Algeria. With land, they were generally able to provide for themselves and to maintain cultures and ecologies which had predated the French – including systems of medicine and healthcare – but without land they found themselves wholly at the mercy of a colonial state whose conception of care was, as we have seen, strictly delimited. Famine was a cause of landlessness as the impoverished moved to the cities, whilst the confiscation of land was a cause of famine as local ecologies were destroyed. Similarly, landlessness and local revolt were also indissolubly bound, for local peoples well understood the shorter- and longer-term consequences of the appropriation of their lands, while that punitive confiscation was also an avowed policy goal of the colons after the defeat of insurrections.

We have seen that Davis's work shows how scientific arguments were deployed in the colony in conjunction with a historical narrative of post-Roman decline to assert that the desertification and impoverishment of the Maghreb might be reversed through the application of progressive imperialism. The problem, of course, with such a project was that its supposedly objective descriptions bore little connection to the environmental reality of the lands of North Africa, where desertification had generally been effectively managed and was only exacerbated through the policies of the colonists themselves. Echoes of the warping of logic found in the colonial psyche identified by Nandy were again apparent here. In Émerit and Yacono's work on the Chélif valley we see, on a micro level, how existing collectivised agricultural economies were destroyed, how large landowners, on the French and local side were encouraged and how smallholdings disappeared, at that very moment of the diminishment of populations in famine and migrations to cities.[91]

The manner in which the destructive effects of political and military decisions on Algerian bodies could be construed as natural and inevitable consequences of civilising change are well apparent in Joost van Vollenhoven's influential *Essai sur le fellah algérien* of 1903.[92] Writing 30 years after the great land grab of the early 1870s and the elimination of a significant portion of the indigenous population, van Vollenhoven believed that in the bodies and minds of the indigènes and the settlers he could see a civilisational narrative, and its inevitable consequences, writ large. Algerians were characterised as 'primitive' and 'lazy', incapable of dealing with change, while 'competition from colons served to underline the faults of the indigènes'.[93] The 'physiognomic…force' of the colon was contrasted with the 'weakness' of the indigène and van Vollenhoven was determined to express the notion that this disparity was a consequence of choices made by Algerians who would not imitate their betters, in spite of the fact that structural, political and economic factors prevented them from doing so.

In a sense one might argue that van Vollenhoven did acknowledge such determining realities in an inadvertent fashion, for he wrote that 'All of the intelligence and practical good sense of the colon derive from his race, but also from…his long struggles against the realities and peoples of Algeria.'[94] Now when he uses the words 'contre…les hommes de cette terre d'Algérie', van Vollenhoven evidently means to imply a generic sense of struggle against a warlike, intractable people, but I think we might now read these words as a form of justification for ultimately eliminationist forms of politics, for to be against a man is to want him dead and that was after all the structuring logic of these narratives of progressive, civilisational change and racial struggle. What is more, the manner in which the writer expressed these ideas was in terms of action and intentionality rather than simply the abnegation of responsibility that could be invoked in citing historical change as an explanation for the waning of the Algerian population.

Van Vollenhoven essentially described the formation of a national market, an Algerian economy, in the later years of the nineteenth century. Up until that point, much of Algeria was not effectively conquered or it was governed by competing colonial forces, but under the Third Republic it became clear that Algeria was now controlled as a single economic space. Evidently, it was the subjugation of land which was the very basis of this rule.

An increasingly efficient colonial administration was, however, confronted by periodic demands that the absolute power of settlers over the indigènes be tempered, most especially at moments of economic crisis, for the combined power of the market and the state in Algeria were now such that there were no redoubts for locals at such times. The huge transfer of land from Africans to Europeans in the decades after the famines of 1868–72 was seen by colonists as a form of economic Darwinism, but this generalised landlessness brought with it security concerns, for it was well understood that the impact of small changes – such as a poor harvest or falling crop prices – were likely to have devastating effects upon Algerians, for their own economic structures were now wholly destroyed and they lived on the margins of systems over which they had no control, heavily indebted and in danger of starvation at any moment. Such insecurity inevitably brought with it the potential for revolts of the dispossessed.

In January 1899, for instance, the prefect of Oran reported to the Conseil supérieur d'Algérie:

> If the indigènes are more likely to commit crimes than Europeans, while it is true that their temperament, their mental state and their religion's lack of respect for the infidel must bear some of the blame, but we must also acknowledge that a deeper cause should be added to these natural moral traits, which is the *struggle for life* [interestingly, the English words are used in the original]. When the life of the indigène is a fight against misery and hunger, no sense of morality or fear of punishment will hold him back. True punishment is accorded by God in the future, whilst hunger is a form of tyranny found in the present. The improvement of security, therefore, should be based less on a more efficient system of the punishment of crime and more on managed and planned systems of assistance.[95]

Such pleas were of course in some way indicative of the kinds of administrative struggles which grew up between the locality and the centre, in which, as we have seen, there was evidently often a greater sense of duty to those indigènes whom one saw with one's own eyes. The prefect of Oran was clearly determined to stave off local rebellion with the support of the central state, but he also articulated a form of French welfarism appropriate to this particular moment. Where French monopoly capitalism of the period 1830–70 had had little concern with the rights of workers, commercial and governmental alliances had built up in the Third Republic around benevolent notions which promoted the idea of welfare as a means of ensuring social peace and increasing long-term economic productivity. So, in some sense, the prefect was asking that such ideas migrate from the metropole to the colony, but this was something of a vain hope (even though echoes of a medical universalism in the colony might be reclaimed) for it was clear that it was in the interest of most colons and commercial interests that the new Algerian proletariat be thoroughly subjugated and that racial logics be deployed as a means of ossifying a form of colonial monopoly capitalism. As Billard, the sociologist and formerly deputy prefect of Orléansville reported, all Arabs, no matter how comfortable their previous existences, seemed destined to 'descend' to the 'proletariat'.[96] Even van Vollenhoven acknowledged, 'the truth was that, given his poverty, the indigène needed all his acuity simply to not die of hunger'.[97]

This is not to doubt the concern of men like the prefect of Oran, for a major part of the life of a colonial administrator was to witness and record the suffering of his subjects, but the outcomes of their pleas tended to be reports and enquiries whose sole outcome was the generation of the investigation itself. Thus, when the doctors at the Orléansville hospital reported that the majority of their patients were malnourished and that famine had become endemic in the Chéliff region, the Governor General demanded an enquiry into this situation.[98] Yet, the conclusions of such works were often like that of the 'Commission for the protection of indigenous property' which acknowledged that local landlessness was an increasing problem, but reported that this was 'a purely local phenomenon'.[99] These local realities were 'regrettable' but it was not possible to 'generalise' on the basis of such instances, while the writers of the report also gave voice to the views of men like M. Boyer-Banse, who reported, no doubt wholly erroneously, that it was in fact the land of colonists that was most at risk at that very moment.[100]

Van Vollenhoven was not without sympathy for Algerians, but such emotions were essentially meaningless in that they were clearly not based on any conviction that structural reforms needed to be imagined akin to those that had remade the Algerian economy and land as a place of a dispossessed proletariat for locals. He does, however, offer us a powerful sense of the ways in which such debates were constructed, not least in his summary of the colon position on such questions:

> To the Fellah of the Tell: leave the fertile plains of this land where there is no longer room for your lazy, primitive culture! The law of progress had dictated that the land now belongs to the better-equipped colon. You need to heed the advice of France, through your beylik, on such matters. There may be some rocky outcrops on which, with the permission of the forestry service, you might raise your goats, but if no such places remain you must accept your unfortunate destiny. As Sabatier remarked, your race is destined to disappear, to suicide![101]

Van Vollenhoven evidently meant to parody such ultra-colon views to some extent, but he remained convinced that ameliorative legislation could provide 'excellent results' which would somehow temper the effects of what he evidently saw was an extremely powerful socio-economic trend towards the relentless impoverishment of the landless indigènes. His importance comes in the manner in which he reveals the moral language used to describe the contours of a contemporary debate which identified connections between land, ecology, economy and health as a major political issue in the life of the colony. While the specific suffering of Algerians which we saw in Ruisseau is generalised here, there is no doubt that the central question which runs through his work relates to the health of the Algerian people and the question as to whether the logical consequences of France's policies was, as Sabatier suggested, the elimination of that people. As we saw at the start of this book, the structural qualities of such realities were revealed again in the great famine of 1921 when the same debates of the early twentieth century, which replayed those of the period 1868–72, needed to be unearthed to imagine possible responses to the horror of another demographic collapse.

5.7 Politics, famine, health

In the complex of events in the period 1868–72, we should not ignore the manner in which the realms of health and politics were often intertwined. We see, for example, both revolts by Algerian tribes driven to penury and starvation by the policies of the colonial administration and the formulation of new administrative strategies with malign impacts on the health of Algerians which were responses to local insurrections. As noted, this was also a defining moment in terms of struggles within colonial hierarchies and positions taken by the army and the colons, especially *vis à vis* locals, were in part tactical calculations based as much on struggles within colonial society as they were any kind of rational response to the plight of the Algerian peoples. Atop this already rather disorganised political realm there lay the confused character of the so-called 'dual administration' which, as we have seen, was a much less efficient and clear-cut set of territorial jurisdictions than its name might suggest.

What I wish to do in this section, therefore, is to describe how a muddled polity had major impacts on the health of Algerians and to show the ways in which the arguments of different camps had destructive effects on the lives of Algerians. This will involve looking at the memoirs and diaries of French military figures from this moment and subsequent decades as they attempted to avenge what they saw as a double defeat in 1871, for not only had they been beaten by the Prussians, but they had also seen much of their power in the colony wane as the colons seized their opportunity at this moment of military weakness. Interestingly, while there is a massive military literature on this period, there are very few texts written in support of the colons, though, as we shall see, there are texts which reflect a central divide within all French colonial classes, between religious and secular imperialists.

One of the few texts written in defence of the colons was the 1871 pamphlet *L'Algérie Assimilée: Étude sur la constitution et la réorganisation de l'Algérie*, penned by an anonymous chef de Bureau Arabe.[102] His chief problems with the colonial state were that the dual administration sanctioned the existence of two governments with wholly different cultures and ideologies, and that the military administration tended to prefer dealing with Algerians as Algerians and to resist the need for an assimilatory culture. This was exacerbated, in the eyes of the author, in the period of the later Second Empire when the *arabophile* tendencies of the imperial court – when Algeria became an 'Arab Kingdom' – played into the hands of the colonial army.[103] The consequence of the existence of these 'two forms of administration which appeared to be cogs within the same machine, but actually constituted two separate machines' was that an enormous amount of valuable investigations and research were undertaken into the problems of Algeria, but these rarely found themselves enacted as policies.[104]

This observation raises an important question, I feel, with regard to the history of health in this period, for if this author is right we might say that one of the reasons for the terrible rates of indigène mortality and suffering was that they were cared for by administrative systems which were more notional than realised, yet I wonder if this is a wholly acceptable explanation. It is the case that an ethical lacuna and a palliative gap existed as a consequence of this governmental situation, but the logic of such claims tends to ignore the issue of colonial agency in such situations. Was this after all not a moment of relentlessly focused practical action in terms of land appropriation, with its terrible consequences of the dispossession of tribal peoples and their ecologies, and their forced migrations to towns like Ruisseau where systems of care were arranged in such a fashion that they saw no duty towards peoples whose sickness was a consequence of the 'turning of the cogs' of some other part of the colonial 'machine'?

The task which most army accounts of the insurrection of 1871 took upon themselves was to explain why the revolt, or, rather, the series of rebellions, had taken place. Louis Rinn's mammoth history – whose bibliography gives some idea of the scale of publications on this period[105] – argued that the uprising was simply the confluence of a series of local struggles which ought not to have been seen as a collective effort based on a sense of religion, race or oppression.[106]

Military accounts were keen to stress their pragmatic reading of events and the approaches they took to them, for they saw a clear contrast between their own practical efficacy and the speculative inefficiency of colon governance which claimed governmental equivalence but tended to have little real grip on Algerian society. Such a view was also well expressed by Leblanc de Prévois, who argued that the events proved the need to abandon civil administration completely and to clearly assign the task of colonial governance to the army.[107] What interests me most about his establishment of this opposition between army and colons, is his suggestion that the 'utopias' of advocates of civilian rule should be countered by reference to 'statistics, death rates'.[108] On one level, this is a simple repetition of an oft-cited antagonism, but I find it to be both curious and powerful that Leblanc de Prévois chooses to hint that a compelling moral case might be developed against civil rule on the basis that such governance was a more dangerous proposition for the indigènes. This is revealing of a structural tension underlying colonial power struggles at this time for, while there were evidently divisions within army circles, it was clear that soldiers were more likely to identify the well-being of locals as a key task of government, whilst colons were viewed as a class whose imagination of the duties of rule included a wilful abnegation of any sort of responsibility towards Algerians.

A similar assertion was made at the beginning of the argument of an anonymous army memoir of 1871 in which the author contended that the very concept of a *Régime civil* was founded on a cynical form of 'confusion'.[109]

What the soldier meant by this was to suggest that the colons did not really seek to govern Algeria for, unlike the army, they had no sense of having any duty towards Algerians. The values of these 'European' colons contrasted with the 'traditions of France' which had been upheld by the army, who protected the indigènes from the bloodthirstiness of the colons.[110] Now, as we have seen, such a stance took a very partial view of the behaviour of the French army in Algeria, for it was the army which had created a culture in which violence was sanctioned on discriminate and indiscriminate bases, but it is of significance to our understanding of this period that soldiers might now attempt to deploy moral arguments about the care of Algerians against the colonial classes. In one sense, what men such as this soldier were describing was the collapse of welfarist colonialism, its idea of medicine and its peculiar and unique culture which we looked at in earlier chapters. As resources were diverted away from the structures established by the army, it was clear that locals would perceive both the waning of actual services of care and that ethos which supported them. Quite why the prospect of this defeat was so terrifying to soldiers is slightly harder to divine. Of course on one level it increased risks to the army for it made the whole of the country less secure, it encouraged revolts, migrations and other social phenomena which the army found hard to control (and soldiers were not shy of pointing out that most of the country, and all its most troublesome areas, would remain in their purview), but we might also legitimately ask whether a strong ethically driven sense of injustice also motivated the reactions of some in the army. After all, it was the case that soldiers such as those we are reading of here specifically identified the health of Algerians as that which would suffer in this shift towards civil governance, and we ought not to forget that welfarist and medicalising ideologies had had a powerful effect within the army's cadres. While they too may ultimately have been as 'utopian' as the visions of rule the army saw in the colons, they formed an important part of the culture of the army in Africa. This sense of duty to specific groups of local charges was of course a powerful underlying current within colonial politics right up until the later years of the War of Independence when army loyalties to their *spahis* were often seen as countermanding their duties to the metropolitan and civil states.

The distinction made by the 'ancien officier' between the French army and European settlers was a common one in texts written by soldiers, for at a moment when they, and significant numbers of their Algerian charges, had been fighting for the nation on the battlefields of Sedan and the Kabyle, they were deeply resentful of the gains being made in Algeria by a class of people who were not yet even, in the main, French nationals. The colons were, according to Louis Serre, 'people without nationality, heritage, patriotism, family or heart' and there was a clear contrast being made by soldiers in such claims between, on the one hand, these rootless scavengers, and the common purity and nationality of both French soldiers and Algerians.[111]

The colons were people who would willingly 'shoot ten indigènes for a bottle of wine or to turn a franc' and we thus begin to see how an unusual alliance was beginning to develop in the colony on what were ultimately ethical lines.[112] Officers like Serre were disgusted by the morals of the colons and while such revulsion may have also entailed a rather forgetful attitude towards the army's actions in the colony, there is of course a continuation of a moral logic which stretched back to Bugeaud, and which was renounced by the colons, whereby such brutality could be countenanced only in the name of truly strategic goals.

The army had now become the champion of 'true liberalism' in Algeria and as such it sought to occupy this position in such a way that it believed that its identification with liberalism within metropolitan political discussions at the founding of the Third Republic would see its power championed against that of the colons. This was of course a fundamental miscalculation for it mistook the character of the incipient Republic which would prove to be less inclined to support the army or liberal ideas in foreign affairs, but it is understandable as to why such views were held, for if an authoritarian empire should tend towards colonial 'liberalism' why should such a trend not be continued when a more democratic regime emerged in France?

To those who proposed a style of self-government in the colony based on the 'English model', the ancien officier again argued against such a move on the basis of a sense of duty to Algerians, contending that such a change 'would be to the detriment of the indigènes. The European-Algerians would be to Algeria what the Spanish were to the peoples of Mexico and Peru'.[113] Now we will go on to see that exterminatory tendencies were alive and well in sections of the colonial army at this time, but we cannot ignore the fact that for rhetorical, political and ethical reasons, some officers were determined not to abandon the conquest through welfare for they knew that the abandonment of that course might lead the colons down the very path of genocide which had been identified, mainly by military colonial theorists, as the chief policy rival of welfarist assimilation from the very earliest days of the colony. France, according to the ancien officier, faced a choice, for she needed to be aware that choosing the route of the colons would be to 'begin a process of violent conquest':

> She needed to decide if the indigènes, who had given so generously of their blood and their gold to France, would be abandoned to the harmful, feudal financial and commercial interests of the colons; if, in return for their loyalty, Algerians were to be condemned to abandon their laws, their society and, ultimately, their religion.[114]

In one of the most interesting and telling rhetorical moves I think one might find in any French text from nineteenth-century Algeria, the author then goes on to compare his present to lessons which might be learned from the past, contending that the Roman failure to acknowledge the natural and social 'laws' of the Maghreb had led to the defeat of an imperial force that had 'become forgetful of the immutable laws of justice'.[115] It was for this historic reason that France needed to 'resist those who preached violence' and to remain true to 'the traditions of her glorious past' in maintaining a 'liberal and Christian form of conquest'.

Now the reason why this passage interests me so much is of course because of the way it meshes the ethical with the historical and it sees such debates as being ultimately sited in the bodies of Algerians. Treating those people well was a just thing to do and it also transpired that history taught that this was an expedient course of action, for empires only succeeded when they actualised their ethics in the manner in which they treated subjugated peoples and allowed for the maintenance of the ecology of their cultures and lands. This final claim is especially interesting, and was presaged in earlier remarks, for it also reflected an opposition to assimilatory politics and the positing of the idea that a form of engagement with Algerians existed outside of exterminatory and assimilationist tendencies.

I would not want to claim that this 'third way' – and its ethical grounding – had somehow suddenly become the dominant ethical motif in military thinking, but we will see in future chapters, that the very existence of such an idea played an important role in the future history of the colony and the history of medicine in particular. This notion was concretised in some ways in the establishment of training programmes for indigenous doctors and we will see that its failure played an important role in the shift towards new modes of disillusionment and resistance towards French rule.

The significance therefore of this ferment of political and ethical ideas about the future of the colony at this time of crisis was not simply of import at that moment but also in the manner in which very significant debates and disputes in the colony were being structured. Texts, arguments, tropes and rhetoric were not simply words on pages but were formative of the shifting development of a political and moral culture being reshaped in fashions that would have major impacts on the lives of colons and indigènes. Ethics is of course essentially the study of the conceptualisation of moral behaviour at moments of choice, and this instant was very much a key one for the army in Africa. I do not want to claim that a universal shift towards ethical norms grounded in notions of justice and autonomy swept across that army, but they certainly became more prevalent and the manner in which they did so was in reaction to a complex set of political circumstances in both metropole and colony.

This process, by which the army came to identify itself around an idea of justice, was of course also connected to events in Europe and in colonial memoirs, such as that of Beauvois, we find a resentment at the 'unjust criticisms' of colonists and those who suggested that Algerian insurrections were fomented by the army as a means of reasserting the need for their control.[116]

In Serre, the causes of the insurrection are again rehearsed in part as a means to engage in a broader moral and political discussion. What is of especial interest about his text is that it contains a series of exchanges between Algerian and French commanders which purport to support the military's reading of events at this moment. In a conversation with General Lallemand, Caïd Ali contended that it would have been illogical for his forces to revolt against a French army, when his own tribesmen were fighting alongside that army in Europe.[117] It was, therefore, only because of three sets of policies of the civil authorities in Algiers and the provisional government in France, that the tribes had revolted.[118] The first factor was the way in which colonial papers were suddenly filled with colon aspirations to dispossess Algerians of their lands and to drive them into the desert. The second was the decision to offer citizenship rights to Jews, whilst Muslims, who fought alongside France, were treated as a 'despised, subaltern cast'. Finally, there was the fact that colonial papers now relentlessly criticised the army and, in the absence of strong voices in defence of the army, Algerians revolted for fear of 'the domination of these men who wished for our ruin and our extermination'.[119]

In General Ducrot's pointedly titled memoir of 1871 *La Vérité sur l'Algérie*, he described Algeria as 'this land which we have loved with all the ardour of youth, which for so long now has been the object of our dreams and our fantasies' and I think in combining such notions with the significance accorded by the army to the views of Caïd Ali reveals something important about the colony at that moment.[120] While it was not the case that eliminationist fantasies had disappeared from the army, it is striking that by the 1870s such notions should have come to seem so strange within military culture – given their prevalence in earlier decades. What had changed in this time? Was it that the army had suffered in Algeria: that it had lost many of its men in the name of 'dreams and fantasies' which increasingly became its own rather than those of the metropole or other colonial forces? Was it that the army had invested so much of its idea of itself in a vision of the colony which, in various assimilationist and non-assimilationist guises, was predicated on a welfarism and a form of cooperative dependency which was so completely alien to the burgeoning colon culture of the coastal cities? Was it that the army had become more used to moral dealings with Algerian tribal leaders than it had with French civil functionaries? Was it the case that the whole concept of a concentricity of care was essentially a reflection of the duties towards Algerian charges now perceived in much of the army, and that the kind of lack of interest in suffering which we find in the reports of many civil administrators reflected this split of the country into two different moral realms, where the consequences for the lives and health of Algerians were so very dependent on their place in the politics of the administration of the colony?

Before moving on to look more closely at the demographic collapse of this period, let us reconsider the exterminatory idea and its importance in colonial politics. We have established that in 1871 Algerians feared a new wave of extermination at the hands of the colons because of the political turmoil and changes in the colony. We have also noted the worries of soldiers such as the ancien officier who feared that in Algeria his countrymen might treat the indigènes in the way in which the Spanish eliminated the peoples of Mexico and Peru. We also find the Peruvian example cited by the pro-colon journalist Paul Bourde in his work on the Biskra insurrection, for Bourde argued that unless much more radical forms of assimilatóry policies were followed (in which 'Arab society would be destroyed'[121]) then the only plausible opposing policy would be a form of social Darwinism in which 'the inferior races would disappear' as was the case in Latin America.[122] Bourde is of interest, I think, because he reveals the capacity both for the continuation of the way in which the eliminationist argument could be deployed as a powerful form of advocacy for other policy options (or, rather, the promotion of a single alternative to elimination) and the evident longevity of the eliminationist idea itself, well beyond the period when the whole of the country was essentially pacified. Yet it was no longer from within the army that the loudest voices in favour of such policies were found, but amongst the colons and the civilian administration. Such arguments, as we have seen, do not always need to be enacted to warp a political environment and to have outcomes, particularly at moments of crisis, which are attributable to the prevalence of the idea that elimination was a morally acceptable policy option.

5.8 On demographic collapse

The chief question I wish to answer in this section is whether there was an Algerian genocide in the period 1867–72. For a number of reasons this question has not tended to be posed, yet, as we have seen, there are compelling comparative reasons for thinking that Algeria would have been an unusual colony if there had not been some form of genocide. What I shall seek to show is that the significance of this time of crises was that a genocide appeared in a starker fashion than had been the case in earlier decades. This was in part due to the complex of events and pressure on resources detailed above, but my contention is that this moment was unique only in the visibility of the scale of suffering in Algeria. There had existed a genocidal culture from the inception of the colony, though this had never been fully appreciated because the compilation and analysis of statistical evidence pertaining to locals was generally seen to be of little significance, and was not thought of critically or in the round.[123] There was also an inherent implausibility to the idea that a beneficent conqueror might also be engaged upon a genocide.

This story in Algeria was simply not as stark as the cases of colonies such as French Equatorial Africa where most indigènes disappeared in a very short space of time, but there is no reason as to why a genocidal culture might not be one where populations rose and fell. For reasons which are partly explained by the revival of military beneficence in the Third Republic, the generalised exterminatory urge of the French in Algeria waned after the demographic collapse of this period, though it evidently had the capacity to re-emerge at moments such as the famine of 1921 (and as Prochaska noted, there were distinct regional variations, with rapidly declining populations in parts of the country). With such talk of famines, suspicions might be raised as to why the term genocide is being deployed to describe periods when it was self-evident that very large numbers of people would have died because of ecological phenomena which had regional and global affects. We have seen that care for the sick and the dying diminished at those moments when it was most needed, but does this in itself constitute a genocide or simply the mark of a brutal regime? What I shall need to do in this section is to show how events in Algeria went well beyond abnegations of responsibility to care, though I shall also suggest that wilful ignorance may constitute a feature of genocide.

The reason why the title of this section refers to 'demographic collapse' is that I want to explore both the statistical evidence of falls in population and also the manner in which writing about such matters tended towards the euphemistic in its conviction that it should shy away from the intentionality found in terms such as 'genocide', for there needed to be an acknowledgement that complex sets of factors led to falling populations. This, of course, ignores the fact that genocides need to be unravelled in a similarly intricate way, for there are but five essential components of genocide: a desire for the death of others, an explanation as to why this would be a good thing, the organisation of killing, its achievement on a mass scale, and, whether obvious or sublimated, guilt in the later pronouncements of its perpetrators.

While all writers agreed that very large numbers of Algerians died at this time, there is frustrating inexactness about the statistical evidence which recorded the scale of their demographic decline. Such problems with the compilation of data reflected a long-standing form of structural inequality and willed ignorance whereby there was a marked contrast between the obsession with collecting information pertaining to the births and deaths of soldiers and settlers, and a progressive scale of lack of statistical interest in Jews, Muslims in towns and the 'tribus indigènes', who made up by far the largest group in the country. As Rey-Goldzeiguer remarked with regard to the question of calculating the scale of casualties in famine, 'The colonial powers had no interest whatsoever in establishing the truth' in such cases.[124]

Such uncertainty could be used as a means of denying the existence of problems and, most particularly, as a way of explaining how individual, witnessed forms of suffering ought to be viewed in isolation and not as indicators of demographic trends and evidence of structural flaws in the administration of the colony. Thus, we find Paul Bourde using the fact that the census of 1872 showed a fall in the national population of 600,000 (following an earlier fall between 1861 and 1866[125]) and that of 1876 a rise of 330,000 as evidence of the failure in the statistical apparatus of the colonial state.[126] This failing was real but it is quite clear that such faults and obfuscation could allow the denial of realities as great as that huge drop in population revealed in 1872. Taithe's recent analysis of deaths in the province of Constantine due to drought, famine, earthquakes and epidemics in 1867–68 calculated that between 130,000 and 450,000 of the 1.4 million inhabitants of the region perished in that short period, yet the gap between these estimates is almost as significant as the scale of the tragedy they describe.[127]

What is generally clear from that demographic evidence which exists for the first four decades of the colony was that the population of indigènes had at least halved in this period.[128] Writing in 1870, Auguste Dupré expressed a frustration with the 'vague and overly general' data which had traditionally been provided by the office of the Governor General, requesting that much more detailed data be provided on particular tribes and regions.[129] Where we do possess such information, it invariably points towards two things: rapid population decline amongst the indigènes and racialised colonial explanations which stress the historic inevitability of such decline.

Boudin, for instance, showed the general trend towards higher rates of indigène mortality over natality in Algerian cities. In Algiers, there were 791 births and 1530 deaths in 1850, and 664 births and 1103 deaths in 1851. In Oran, there were 1319 births in 1850 and 1276 deaths, with 656 births and 356 deaths the following year, whilst in Constantine there were 773 births in 1850, with 1386 deaths, and 1119 births and 1379 deaths in 1851.[130] Yet, like Dupré 20 years later, he was to express dissatisfaction with the lack of official interest in such trends and research into their causes. Why, he asked, had such a 'social trend of this level of significance not become the object of a special investigation on the part of the colonial administration?'[131] Part of the answer to his question came I think in his subsequent speculation on the possible reasons for this demographic shift. Was it the case, he asked, that urban Muslims were dying disproportionately as a result of impoverishment and the destruction of their will? Was it because of the ending of unions between local women and Turkish soldiers, or was it due to 'that mysterious law whereby certain inferior races are destined to disappear on contact with superior races?'[132]

Prochaska's study of Bône in the period after 1870 in fact revealed that amongst Algerians the overall demographic recovery following the low points of the middle of the nineteenth century masked huge disparities between the death rates of different groups and regions. In Bône, while the European death rate was nineteen per thousand in 1911 and fifteen per thousand in 1926, the Algerian Muslim figure was on average forty-six per thousand between 1915 and 1925, with a high of seventy-three in 1918.[133] Prochaska went on to say that, 'As we might expect, Jewish birth and death rates occupy an intermediate position between those of the Europeans and the Algerians' which rather tellingly reveals the blunt realities of racial politics in Algeria.[134]

It now seems fairly obvious that Boudin's rationalisation of demographic decline – as an inevitable consequence of that 'mysterious law' which governed encounters between superior and inferior races – formed a part of the reason as to why there was a lack of statistical and analytic interest in tracking the declining population of Algerians. After all, if that was an outcome which one would expect to take place, it could essentially be treated as a form of historical or scientific proof and greater energy could be devoted to forming the narrative description of such theoretical speculation than to recording the actual deaths of Algerians individually and collectively.

Chief amongst such theorists was of course Eugène Bodichon who, as we saw, wrote so approvingly of local population decline and who provided a medical and scientific set of policies to accelerate that decline which complemented Bugeaud's theorisation of the *razzia* and an exterminatory politics. The long-lasting character of such ideas was seen when the 1881 census showed a reverse in the decline of the local population leading Battandier and Trabut, two professors at the University of Algiers School of Medicine, to argue that such a rise could only be attributable to an improvement in census sampling techniques, for the underlying demographic truth was that the indigène's 'traditional laziness will condemn him to disappear sooner or later before the more active races'.[135]

Doctors were of course uniquely placed to offer both qualitative and local quantitative pictures of demographic decline. In 1864, for instance, Warnier wrote an article for the Algerian newspaper *Courrier de l'Algérie* entitled 'The human cost of Algeria'.[136] The chief point of his piece was to note that while the European population in Algeria was growing fairly rapidly, the local population was declining in a steady and regular fashion. The first of these trends was one which he was keen to impress upon readers in France who still believed Algeria to be a very dangerous place to settle, but his chief point was that a major population shift was taking place in Algeria.[137] He noted that:

> If a sense of equilibrium is not returned to the tribes, in terms of an increase in the birth rate (which seems unlikely given that the local population has been stable, if not stagnating, for centuries), it would be easy to predict that the local Muslim would become the exception in Algeria. The European population would soon become a majority force and eventually outnumber locals by a factor of ten.[138]

Warnier's article is of interest to us for the manner in which it demonstrates that there was a clear awareness amongst colons, and doctors in particular, that the French were inducing demographic change in North Africa (even if writers like Warnier alleged that Arab depopulation had been occurring for centuries). One can see why Rey-Goldzeiguer singles out Warnier as the exemplary leader of that colon mentality which convinced itself that its ignorance was in fact knowledge, whose thought was utterly directed towards the justification of conquest and dominance.[139]

In the period of crisis being studied, what we are especially interested in is that combination of rapid demographic decline with expressions of desire or thanks for such trends, such as admiral Gueydon's hope that the arrival of new colonists in 1871 would lead to the 'destruction or servitude of local peoples' or General Lapaset's remark of the same year that land confiscations would create 'an abyss which will be filled one day with corpses'.[140]

The best-known and most comprehensive demographic study of this period, Dr René Ricoux's *La Démographie figurée de l'Algérie*, exemplifies the attitudes set out above.[141] Locals were generally ignored in Ricoux's massive study which displayed an obsession with the compilation of data – looking for instance at deaths by season, deaths at different climatic moments and deaths amongst European nationalities – whilst evading any form of detailed analysis of the chief demographic trend in nineteenth-century Algerian society. Such investigations were unnecessary because while European deaths merited detailed scientific scrutiny to generate understanding and policies, a simpler form of racial and historical science explained falls in the population of the indigènes. These were, after all, 'inferior and degenerate races' whose current predicament, when contrasted with the glories of their past, served to prove the inevitability of their 'rapid and ordered disappearance'.[142]

When the census of 1872 revealed that the Algerian population had fallen from 2,652,072 in 1866 to 2,125,051, the story which this confirmed to Ricoux was one of natural wastage, with little thought to the ambiguity inherent in this claim's scientific pretensions actually being founded on ideas of destiny more appropriate to a pre-scientific moment (or indeed to those fatalistic and religious qualities invariably attributed to Algerian Muslims by men like Ricoux).

Like Battandier and Trabut, and in spite of his great faith in statistical methods across his work, Ricoux could not accept evidence of a rise in population (to 2,462,936) in the census of 1876, for that data undermined the racial theory on which his very limited engagement with local demography was based. He suggested that the census had been based on faulty data sets and alleged that his own researches showed evidence of further falls in local populations. Far from rising again in numbers, local populations were destined to 'inevitable disappearance'.[143]

Interestingly, Ricoux also perceived a need to stress that the French had had no part to play in this demographic collapse; in particular policies of 'refoulement' and other examples of the brilliantly euphemised 'politique humaine' could not be said to have contributed to Algerian degeneration.[144] The manner in which local populations did rise and fall, and the fact that when they fell they tended to do so in a dizzying fashion which was not especially amenable to scientific explanation, forced Ricoux further down paths of historical explanation. Approvingly citing Vinet, he agreed that historical 'laws', such as those seen in South America, determined the fates of backward peoples on their encounters with superior races.[145]

In spite of his certainty that France and her humanitarian politics bore no responsibility for Algeria's falling population, Ricoux's own expositions on this subject can be seen to reveal his own faith in historical explanations, as we see in the following remarks:

> What is killing the Arab population is that interchange which is taking place between her fatalism and the initiative and organisation of her interlocutors, who are open to exchange. Yet the Arabs refuse to adopt the habits, procedures and institutions which are necessary for such exchanges, and as their dealings with Europeans increase, so does the incidence of epidemics in their populations, which they fail to check through the adoption of the rules of hygiene, diet and medicine proscribed by modern science.[146]

In other words, the infantilised civilisational state of the Arabs was such that they could not see how a failure to fully assimilate with their conquerors would necessarily lead to their own destruction. Such a medicalised account of historical inevitability was of course very different to the paternal benefi-cence of the idea of the medicalised state, which is wholly absent from Ricoux's explanatory model, for the very reason that that which he desires cannot be obtained if the responsibility for the playing out of history is seen to be shared by the French and Algerians in his demographic model.

He went on to assert that 'The Arab population is dying out as a conse-quence of her commercial relations with the civilised world', which was true in the sense that local economic structures were destroyed and lands confis-cated, though Ricoux would evidently not wish French agency to be high-lighted in this explanatory model. He then added that any possibility that Algerians might be somehow saved through intermarriage with Europeans was a clear impossibility, for 'the French would have no interest in sacrific-ing their natural qualities and moral superiority' in admixing with a 'cor-rupted race' whose 'tainted blood' bore within it the traits of 'dirtiness, bad faith, criminality and every conceivable form of moral and physical depravity'.[147] He recorded with some pleasure that at the time at which he wrote there had been only 120 marriages between French citizens and Algerian Muslims.[148]

The significance of the extreme quality of Ricoux's 'repulsion' at the idea of any form of racial mixing in Algeria is of course evidence of the manner in which Algerians were textually dehumanised to the point where even those branches of science devoted to recording the character of life and death – as was the case with demography – constructed a pseudo-scientific edifice of moral hatred for the *indigènes*. My own view is that authors such as Ricoux reveal the way in which the delusional and destructive qualities of French imperialism in Algeria, and the place that medicine played in this, have tended to be underestimated, or somehow occluded by a concern with the medicalising project.

In Ricoux's eyes, disease and degeneracy were forms of moral judgement on a people, and his work stands as an exemplar of the anti-scientific character of French colonial writing, whose mask of enlightened science only seemed meaningful in a very particular culture where its *mélange* of ethical judgements, pseudo-scientific ethnographic theorising, imagined historical determinism and loathing could be construed as possessing an internal logic. Removed from that culture, the purpose and the structure of his system are revealed to be variants of the whole racial-medical-moral complex of ideas which we have seen as much in Bugeaud as we have in the texts of doctors like Bodichon. Such writing acted as a kind of wish fulfilment in the manner in which it not only sanctioned the ignoring of demographic collapse, but also celebrated it as evidence of the playing out of some deeper set of structures in global history. If we think back to the Ruisseau correspondent we begin to realise how aberrant his instinctive sympathy for the indigènes was in such an environment, where the 'politique humaine' of *refoulement*, land confiscation and the *razzia* could be denied to harm Algerians, even as many of the architects of such strategies had, like the writer from Ruisseau, to see the bodies of those who suffered and died as a result of those policies which could not themselves be admitted as causes of those agonies. Le Cour Grandmaison suggested that the 1880s saw the configuration of a new form of 'racialised, moral, hygienist' discourse, but it seems evident to me that that potent combination was written into imperial culture in Algeria from the earliest days of the colony.[149]

What then might we say about the failure of men like Ricoux not only to attend to Algerian suffering, but also to celebrate it in their theorisations of its inevitability? On one level, I believe the complex theoretical schema of men like Ricoux were a form of overcompensation for what was in essence a set of blunt and obvious facts: Algerians were dying out as a result of French policies in the colony, they were dying at a rate greater than they had under the Turks or in earlier centuries, and there was evidently a connection between French desire that this should be the case and it being the case.

The writing of men like Ricoux also allowed for the possibility that something akin to genocide could be construed as, at worst, a form of historical inevitability with which man could not tamper, and, at best, a kind of counter-intuitive enlightenment as the pace of historical progress was accelerated by those in its vanguard. This was in a sense an informal genocide as men who described themselves as good stood by and did nothing, but its informality (by which I also mean to indicate the manner in which it was spread across time and places, resisting counting and classification) should not occlude the fact that it was also a medicalised and theorised enterprise. In this sense it was far from accidental since a theory was postulated – that the native classes were destined to extinction – and those actions and policies which advanced the actualisation of that theory were celebrated and added to scientific literatures on the colony.

Theorisation was also, I would suggest, a form of assuaging of guilt for it served as a kind of distraction from the apperception of reality. It posited some higher, abstract realm of truth which allowed the suffering of others to be bracketed in the name of that higher order of reality, but, as the Ruisseau correspondent showed, the edges of that theoretical culture might easily be punctured at moments of crisis. The risk which was then run was that the world may seem as if turned upside down: that which was explicitly described as a moral good in theory could be seen to be a form of maleficence in practice; that which was seen through a dispassionate lens of exactness was revealed as an uncomfortably rough, human story of pain in practice, while the putative palliative effects of a medicalised system were seen to have another side in the manner in which medical science was used as a key part of the edifice of ideas which justified a genocide.

In order that this psychological system be maintained, it was imperative that the sympathy displayed by men such as the Ruisseau correspondent was countered by accounts which identified the playing out of racial theories in practice, even at moments of great suffering. A good example of this rhetorical tactic can be seen in Dupré's letters, as in his description of Algerians' response to their plight in 1870:

> Following those precepts which are bound to his nature and his instinctive laziness, the Arab works as little as he can, just enough to get by. Even when confronted with destitution and hunger the Arab remains lazy, preferring death to the slightest sacrifice of effort. The recent famine in Algeria strikingly proved this point. Whilst the workhouses [chantiers] were opened to them, many stayed only to work a few hours before leaving, in spite of the fact that there was no way that they would find food outside those places.[150]

In other words, Dupré too believed that events such as famines could be viewed as being socially constructed rather than natural disasters, but the difference of his account to that of Davis is that here the Arab played a suicidal role in which he was programmed to choose death over work. There were of course both practical and moral reasons as to why Algerians might choose to abandon French workhouses at this moment, for they were evidently centres for the transmission of disease and they were places where both dignity and autonomy were denied, yet Dupré needed to insist that only racial destiny could explain the actions of those who failed to see the offer of freedom inherent in work at the camps. We might also add that some Algerians would have left camps due to their being physically unable to work given the weakened condition in which they arrived at such places.

Dupré's words show the sense in which the French believed that there was some form of guilt to be assuaged at this terrible moment when so many were dying around them, but a combination of three factors demonstrated that the French as a group bore no responsibility for these things. First, the workhouses demonstrated the continuation of the beneficent welfarist tradition, and, what is more, such places combined medical aid with a pedagogic intent, for even whilst healing the sick they would orient them towards a better life through work. Second, a considerable degree of blame could be ascribed to natural events for it was 'the famine' which 'strikingly' proved 'a point'. Third, Arabs often chose death, and the resonance of this observation was founded on a massive body of thought about Algerians and their ethics which had insisted, since the 1830s, that Muslims saw death as a form of destiny and that their fatalistic tendencies held them back from a true appreciation of the meaning and value of life.

In such difficult and intractable circumstances, the French did what they could but both parties in this dialogue were evidently aware that higher cultural and historical forces were at work which would ensure that any palliative efforts on the part of the French would necessarily be countered by the racial preference of Arabs for self-destruction. Such theorisations and the edifice of moral and historical claims which underpinned them always of course ignored actual realities and more readily apparent social and political explanations for sanitary and demographic catastrophes in the colony. Yet I believe that even writers like Dupré shared some similarities with men such as the Ruisseau correspondent, for Dupré could not wholly ignore the realities of the things which he saw and felt duty bound to offer some description of the horror of these situations of workhouses, destitution and hunger. Such writing needed to be accompanied with claims of rigid moral certainty of a kind that begin to offer us access to that complex of ethical claims which assured a particular kind of domination in the Algerian colony.

5.9   Conclusion: on care and killing

Across this chapter I have sought to show that the terrible events of the period 1868–72 were by no means aberrant in terms of the colonial history of Algeria; rather, we should see them as stark proofs of the moral frames with which the French built their relationships with Algerians and their behaviour towards them. Of course, this moment was significant in a political sense in that it saw a shift in power between the army and the colons, but there has been a tendency to overstate the novelty of the violence which the colons directed towards Algerians after 1870. It had after all long been the case that even the putatively universalist medicalising state was founded on a set of racial distinctions and pictures of a duty of care which tended to exclude most Algerians from the beneficence of this project. What is more, the medicalised state, in terms of its ideas and practitioners, also formed a central part of a complex of scientific, historical and ethical theories, which, at best, stressed the utter difference and subhuman character of Algerians and, at worst, offered justifications for the imagined and actual extermination of significant portions of the Algerian population.

The severity of the concatenation of crises at this time provided a series of opportunities to the French to destroy local ecologies, economies and peoples, for which there already existed a powerful set of moral justifications which had been developed in earlier decades. Mike Davis's argument that famines were essentially human constructions which tended to serve a convenient genocidal purpose is borne out in this Algerian example, though the complexity of the politics of the colony in the last days of the Second Empire and the first years of the Third Republic has tended to occlude this fact (I have understated the complexity of Algerian politics in the period in this chapter, most especially by omitting details of foreign policy debates and discussions of Algeria and pan-Islamic and Mediterranean questions). Rey-Goldzeiguer was surely right when she argued that the series of famines and crises of this moment presented colonial forces with a perfect opportunity to destroy their enemies; what is more this could often be achieved through the destruction of forms of social, economic or cultural power, for the directness of the *razzia* could be avoided whilst achieving the same ends (within a culture that putatively cared for those being eliminated).[151] I do not wholly agree with Rey-Goldzeiguer's characterisation of this time as being one in which 'Un monde se disparaît, un autre se prépare', since I wish to stress continuity where she sees a structural shift, but there is no doubting the force of the argument which lies behind the elegance of her words.[152]

Memories of beneficent intent began to fade in the context of this brutal and confusing political milieu which served to emphasise the failure of the French to have truly conquered Algeria after four decades of occupation, as we see here in the words of a prefect to the director of the hôpital des indigènes at St Cyprien des Attafs (the centre of the revival of Christianity in the Maghreb, we will remember):

> Amongst the elderly and the incurable, there are always certain individuals who, on account of their age, infirmity or their characters, greatly pain the staff of the hospitals because they are the cause of a great deal of trouble and disorder. Nevertheless, we are obliged to keep them here because these unfortunate dregs of society would not be able to find shelter outside of these hospitals that were expressly created for them.[153]

France therefore bore a duty which it should like to shake off for such recalcitrant patients revealed not only their own undeserving natures but the essentially worthless character of the cultures from which they came. That such cultures had had established systems of care before the arrival of the French, and that such structures were systematically destroyed by the French in their construction of Algeria, was wholly unapparent to writers such as this prefect.

A similar form of resentment could be found in the manner in which the utter difference of races was given the lie by the fact that diseases such as typhus saw no distinction between Algerian and French bodies, though such movements of sickness could be used as a means of justifying still stricter forms of quarantines and the separation of Europeans and Algerians. And yet the spread of epidemic disease also contained within it traces of the project of beneficent medicine, as at the hospital of Fort l'Empereur where thousands of indigènes had gathered. There, all the staff at the hospital contracted typhus, leading to the deaths of at least twelve military doctors and fifteen nurses.[154]

Such exceptions do not however reflect the general position of French medicine in these crises. While medicine relates to questions of life and death, it is not bound to preserve life at all costs, for medicine and the ethical systems which underpin it, offers justifications for death and suffering as well as healing and saving life. Most colonial doctors and administrators saw the trauma of the period 1868–72 as a form of historical proof which justified racialised systems of rationing. Such ideas necessarily tended to induce demographic collapse amongst indigenous populations, which was not simply a cost of rationing but also a benefit, for a deeper understanding of the structures of this situation revealed that the health of the new Algerian body politic would be much enhanced through such purging. Seemingly maleficent outcomes could perversely prove beneficent intentions, for while we do find cases such as that of the Ruisseau correspondent where the horror of individual suffering induced calls for a just response, to most in the colony such ideas implied ideas of autonomy and equal personhood that could not be countenanced.

Later crises, such as the famines of 1920–21, which like those of 1868–72 were also accompanied by waves of epidemic disease, reveal a very similar cultural *milieu*. While Julien and Nouschi were able to identify structural changes in Algerian society and economics induced by France which played leading roles in the construction of famine, contemporary colonial responses revealed the way in which Algerians were judged, despised and feared in equal measures.[155] At the start of this book we read of the mayor of Cherchell who wrote to his prefect in April 1921 to demand help in holding back the 'invasion' of the sick who 'menaced' the gates of his town, threatening it with typhus, and in May of that year he stressed the danger that such people posed to 'public hygiene', implicitly acknowledging the manner in which Algerians were excluded from notions of the 'public' for all of the reasons set out in this chapter.[156] These were people who needed to be 'purged' from the towns, for they had come there 'not to work and to participate in the life of the commune, but to exploit the system of public charity'.[157] Here of course we saw an exact replica of forms of argumentation from 1868–72, in terms of the moral stress on work, on the local obviation of responsibility for structural factors which had led to such situations and the determination to frame all such discussions in a language of morality and judgement, in part as a means of providing an ethical grounding for knowledge that the failure to provide care in such circumstances was essentially a decision made in favour of the acceptance of the death and suffering of others. At both moments, it was the behaviour of administrators as much as doctors which determined the culture of health in the colony.

When the French had arrived in Algeria in the 1830s, a key component of the historical justification of their conquest had been their assertion that French medicine and welfare would offer hope to a country that had often been ravaged by famines. By the 1860s, such rhetoric had disappeared. Was this because a fundamental shift had taken place from the moment of a medicalised humanitarian imperialism of the early decades to a more brutal set of realities once Algeria had been pacified? My contention would be that this was not the case for the main thrusts of the moral conquest of Algeria had always been towards destruction and the elimination of local peoples, with such discussion in some sense morally assured by layers of beneficent welfarism. To return to my beginning, the period 1868–72 does reveal that answers to the question 'For whom should we care?' had narrowed for the acceleration of those patterns of mortality which had seen the indigenous population fall by at least a half since the arrival of the French, was now seen not only as an opportunity but also as a moral good to those in possession of a knowledge of ethics, destiny and history.

6

On the Just and Sovereign Testimony of Abdel Kader ben Zahra

6.1 Abdel Kader ben Zahra's letter to the Governor General

I write, Monsieur Governor, as a man who holds only the position of health officer, but I hope you will realise the great effort it has taken me to reach even that rank. Out of a population of three million Algerian Arabs there are just four colonial doctors, two health officers and two other medical officials. Those appointed as doctors have qualifications no different to my own – simply secondary education without a bachelor's degree – but they had the great fortune to work under Governor Chanzy, who felt moved to alter their ranks from health officers to doctors. In those days, the authorities never missed the opportunity to encourage young Algerians who wished to study and to assimilate themselves with their French masters; to enter into the new life which France offered them. I would like to believe that such policies continue to this day and that your office will support the development of more Algerian doctors so as to encourage our country down the road of civilisation and progress. ... Today, however, with my health officer's diploma I find myself without resources or a route of advancement. I feel like a sea-going traveller who, when his ship reaches the middle of the ocean, is told that he must disembark for he will be taken no further. Because of this, Monsieur Governor, I feel that I would have preferred to have not travelled on that journey and instead stayed on my own shore, where I would have been content with my savage existence and not troubled by these brick walls and obstructions which now face me.[1]

6.2 An Algerian's experience of the idea of medical imperialism

This chapter offers the first account of Algerians' experience of the idea of medical imperialism and its realities. Along with the following chapter, it relies heavily on the writing of Algerians working in the colonial medical administration and the analysis of hitherto unstudied personnel files held at the Centre des Archives d'Outre Mer. Based around the study of a single character, this chapter re-examines a set of debates explored throughout this book, from the discussion of the creation and cultural dominance of an idea of medicine found in Chapter 2, along with the limits of humanitarianism and the realities of the translation of the idea into practice, as well as discussions of exterminatory urges and the consequences for health of the crises in the period 1868–72 described in Chapters 4 and 5.

The chapter focuses on the career of Abdel Kader ben Zahra and, most especially, the letter of May 1884 cited above. Drawing on writing from across his career, the chapter investigates the relationship between the letter and ethical discussions of both autonomy and justice. It closes with a consideration of the environment in which he worked at Tuggurth, which leads us towards the following chapter's consideration of the careers of a number of ben Zahra's Algerian-born medical contemporaries, many of whom also served in Tuggurth.

The central thesis of these two chapters is that the careers of ben Zahra and his colleagues act, first, as proofs of France's failure to implement the idea of medical imperialism and, second, of the manner in which Algerians were able to draw upon ethically driven forms of critique as a means of expressing their disillusionment with the manner in which France reneged on her beneficent medical promise. Ben Zahra's writing is, I believe, comparable in its acuity to the work of Hamdan Khodja, while both writers lie within a tradition of forms of moral resistance to French rule. This struggle began in the earliest days of the colony with men such as Khodja, before continuing through the nineteenth century with writers like ben Zahra, and then it fed into the rhetoric of thinkers such as Ben Bella and Fanon in the closing days of empire. Like Khodja, ben Zahra was initially prepared to trust the French, but unlike many Frenchmen in the metropole and the colony, he was not simply satisfied with the articulation of ideas of beneficence, for he came to see the forms of moral duplicity inherent in such claims to goodness as a wholly different, more brutal and uncoordinated world formed around him.

In the writings and stories of men such as ben Zahra, we therefore find precursors to the thought of Fanon and the Front de Libération Nationale, and their conviction that opposition to France should be based upon a recognition of the malign contribution of the colonial power to the health of the Algerian nation. This took the idea of medical imperialism full circle in reappropriating it as a form of opposition to France, diametrically opposed to those initial projections of its seductive, conquering powers.

The study of the means by which scientific thought and debates about science served as a means of contesting imperial rule has become an important theme in recent writing in the history of medicine, most importantly, in this context, in Clancy-Smith's description of the power and independence of a centre of Islamic scientific education in the Algerian Sahara at this time.[2] Complementing Clancy-Smith, these chapters mark a move in studies of nineteenth-century Algeria towards the articulation of Algerian voices, which evidently also connects to the approaches of historians of colonial medicine such as Arnold, Andrews, Cunningham and Cooter, who have insisted on the importance of work which views medical encounters from the perspectives of those whose lives were medicalised by Europeans.

It is important, therefore, that we listen to both Algerian and French voices from the nineteenth century – to ben Zahra and Khodja as well as Bodichon and Bertherand – for what we can then see in the history of medicine is an encounter which was conceived of in moral terms and which we now ought to try to capture in ethical language. Part of the value of this form of ethical reconstruction is that it takes us to a conclusion which we might not otherwise reach, which is that the project of medical imperialism failed not simply because it was, in some senses, a delusional idea which could never be as fulfilled in practice as it was in writing, but because the concepts of empire, the civilisational conquest and medical imperialism, quickly lost their seductive sheen when they were considered by people who operated with more complex notions of ethics than their new colonial masters. It is, as we have seen, a somewhat telling form of Western cultural arrogance which believed concentrations on ethics of justice and autonomy to be modern innovations which became significant only after the Second World War, when, to take just one example, such notions had played important roles in the ethical thought of the Maghreb for more than a millennia.

In nineteenth-century Algeria, the idea of colonial medicine was so bluntly focused on a narrow basis of assured beneficence that its practitioners and the architects of systems of health were never able or willing to adapt to an environment in which a commitment to goodness had never been considered sufficient ethical justification in and of itself. Ben Zahra is of especial interest to us because he was brought up as a living embodiment of the idea of medical imperialism, and yet even he came to reject its claims as being both false and immoral.

In the letter of May 1884 cited above, he wrote to the governor that he had only learned to read and write from the age of ten, when he had been recruited to the army, coming from a rural background where he had never before seen a European. His training there had concentrated on military duties and gymnastics and it was only through his great fortune in being selected for one of two scholarships to the Lycée d'Alger by Chanzy at the age of 13 that he had had the opportunity to become formally educated. There he worked extremely hard, in spite of the fact that some options, such as the study of Latin, were ruled out for him because of his late start in education. At the age of 17, he had won a bursary to the medical school in Algiers and was well aware that his career up until that point had been led by a great deal of luck and his colossal drive to succeed when opportunities were passed to him. He rightly saw himself as a great rarity amongst his people, but on encountering the realities of colonial medicine, it was to those people and his 'own shore' that he wished to return, for a life that had begun in hope closed with a sense of despair and betrayal.

6.3   Medicine and autonomy

In medical ethical literatures, the concept of autonomy is generally discussed with regard to patients and their rights. Such literatures tend to see a valorisation of patient autonomy, and the articulation of medical choices to patients, as being a recent phenomenon, relatively unknown in pre-modern medical cultures where patient autonomy was incompatible with professionals' monopolisation of knowledge and power. There are, I think, reasons for believing that the innovative qualities of the concept of patient autonomy are somewhat overstated, even if realistic, if western Europe is taken as a global arbiter in the history of medicine, but here I wish to focus on a slightly different account of autonomy: that of the medical professional to act as a force for good in as unhindered a fashion as possible in her or his work. My reason for so doing is that ben Zahra's disillusionment with his life, and what he had become, seemed heavily dependent on the idea that he had originally believed that he was being schooled in a way of life which would allow him to express his identity through medicine, whereas he came to discover that in practice this nascent identity was continually challenged by forces and structures outside his control.

Ben Zahra's reaction to the exigencies of his professional life, and the denial of a promise of selfhood, drew on a broader Algerian experience which, I would suggest, had its roots in Arab-Islamic concepts of health and ethics. Whilst ben Zahra grew up in a European environment from a relatively young age, he was evidently deeply familiar with his own cultural heritage, not least through the visits he made to patients. To his clients, a doctor, in the Islamic tradition, was a figure who stood outside of many conventional, social and political hierarchies, whose skills were, for some, a gift from God (and who was performing a religious duty in the act of healing others), while for others the doctor stood as a representative of secular scientific endeavour had traditionally coexisted with religious views of health within Arab-Islamic culture. In either case, the moral charge which doctors took on when accepting their calling was a heavy one, for they had duties not simply to act justly towards individuals but also towards communities. Additionally, they were obliged to ensure that their patients saw questions of sickness and health as being related to the good life and to the relationship between the individual, the world, and, to some, their God. This was not as reducible as French colonists believed to the notions that illness was seen as divine punishment, nor that Algerians believed that cures could be found in religious medicine or improvements in morality, for one of the central mistakes made by the French in Algeria was to underestimate the complexity or medical practices and the heritages in which they were grounded.

Ben Zahra's letter cited above can be seen as an antecedent to Fanon's note of resignation from the hospital at Blida-Joinville, where he wrote that, 'If psychiatry is the medical technique that aims to enable man no longer to be a stranger to his environment, I owe it to myself to affirm that the Arab, permanently an alien in his own country, lives in a state of absolute depersonalization.' Fanon argued that the nub of French colonialism had been the manner in which it had depersonalised Algerians; that it had denied their existence as autonomous human agents. Algerians had been made 'aliens' in their own country in very many ways, but one of the most significant of those was the manner in which that tiny number of Algerian professionals who were promoted to positions akin to those conventionally reserved for Europeans, invariably found that the ultimate point of such exercises was to deny their autonomy, their power to act in any kind of independent fashion and for the final strands of the beneficent arguments of civilisational imperialism to be extinguished on a local level by the superiors, rivals and clients of doctors like those who served in Tuggurth.

We see this in ben Zahra's letters to the Governor General in which he lamented the fact that he found himself caught up in a struggle between rival forms of administrative authority, which was at its most tense in towns like Tuggurth, which lay on the edge of the 'civilised' portion of French Algeria, where the army was committed to preventing the encroachment of civil powers. As ben Zahra said in one of his letters, had not all his problems arisen 'because of the fact that I am a civil servant'?[3] A typical example of being caught up in such struggles was seen in regard to his access to the local pharmacy. This institution was controlled by the local Bureau Arabe and the lieutenant in charge would frequently frustrate ben Zahra's work by refusing to allow him entry to the pharmacy, preventing him from having his own key to the building and failing to reorder essential requirements for the town. Ben Zahra felt understandably emasculated for he was not able to fulfil one of his chief roles in the town, which was the provision of 'free care to all the sick, irrespective of their wealth'. Echoes of his Islamic conceptualisation of his duty as a doctor are readily apparent in his words here, as is the need to develop a more nuanced picture of the army and the history of health than that presented in the previous chapter. While it is true that the military authorities on the edges of Algeria had essentially abandoned the project of subjugation founded on hatred, just as the civil authorities of the coast became more seduced by this idea, ben Zahra's career showed that the army's engagement with Algerians and Algerian culture, while less overtly brutal than had been the case in the past, nevertheless focused strictly on operational objectives rather than the development of a new cultural understanding with locals.

Furthermore, ben Zahra had been assaulted by a certain Commandant Schérer and in ben Zahra's account, this had amounted to a concerted effort to diminish his authority in the town, for Schérer had chosen to attack him in front of the Agha and his entourage, whilst a number of locals waited for ben Zahra's assistance at the pharmacy.[4] After the attack, Schérer had driven away ben Zahra's clients. He was convinced that the reason why Schérer had behaved in this fashion was because ben Zahra had delayed handing back the key to the pharmacy as he had had to travel some distance to a patient in need. Ben Zahra wrote to the Governor General to say that he felt sufficiently threatened by the actions of Schérer and other officers that he had taken to carrying a revolver with him at all times.[5]

Other forms of habitual intimidation included the fact that Schérer and other officers delighted in releasing patients from the hospital before their discharge had been authorised by ben Zahra. Similarly, Schérer would refuse to countenance the release of patients whose discharge had been authorised by ben Zahra.[6] As ben Zahra explained to the Governor General, 'to work with the Arabs it is necessary to possess a certain amount of prestige' and it was this which the army were determined to ensure was eliminated in his case.[7] Army officers would routinely dump manure outside ben Zahra's house to diminish his status in the town, and refused to stop doing so in spite of ben Zahra's protest that this presented a danger to public health. The character of the animosity which the army felt towards a doctor who was not only a representative of the civil government of Algiers but also a representative of a race so despised by many in the army was made plain when Schérer announced to ben Zahra, 'You delight in playing the big man [le grand seigneur] here, but it doesn't wash with me.' In other words, the state may have led you to believe that you were in the process of becoming an autonomous professional and a source of power and expertise, rather than simply its subject, but the global and local reality of life in the colony will prevent this coming to pass.

6.4   Medicine and justice

As well as displaying a sense of anger at the denial of autonomy which the medical life provided him and his patients, ben Zahra's writing also displays a strong sense of injustice at the manner in which he had been treated. The behaviour of men such as Schérer offers a partial explanation for this, but to fully understand ben Zahra's ideas about justice and injustice, we need to look more closely at the words of the letter which began this chapter.

As in the case of autonomy, modern medical ethics has tended to assert that questions of justice were either insignificant or absent from most cultures before the modern bioethics revolution. Aside, therefore, from allowing us to understand how a particular moral culture of health functioned in nineteenth-century Algeria, ben Zahra's writings also add to and challenge aspects of the history of medical ethics. They also reveal one of the origins of the ethical optic through which I have been trying to view the story of medical imperialism in earlier chapters, for across this book my goal has been to develop a critique which borrows from the original critics of the colonial idea of medicine in nineteenth-century Algeria.

In his letter of May 1884, there are evidently connections between ben Zahra's sense of being treated unjustly and the denial of his autonomy. He had felt the life he had been given by the French as a form of privilege in which, through work, luck and destiny, he would eventually be assimilated into the community of his 'French masters'. He would thus enjoy a 'new life' in which his Algerian, Arab and Muslim identities disappeared in his own eyes and in the eyes of others, and in this sense the dream he subscribed to was one of an abnegation of autonomy (with the concomitant forging of a new autonomous identity as a French citizen). It is demonstrative of the strength of ideas of civilisational imperialism and colonial medicine that men such as ben Zahra should have been able to arrive at such a view of the world. Yet, as we read, this 'road of civilisation and progress', ended for ben Zahra in what seemed like an arbitrary fashion, for further chapters in his journey towards becoming a doctor and a Frenchman would not be written for reasons of colonial politics and economics. In this sense, his career was utterly emblematic of that gap between the idea of medicine and its implementation. It was also revealing of the shallow character of the assimilatory impulse in the Algerian colony, whose rhetorical valorisation has led it to be overstated as a French policy goal I would suggest.

More than with any of his other correspondence, there is a sense in which the letter of May 1884 was written as a self-conscious discussion of ethics and morality, for it represented and described a moment of choice and realisation in ben Zahra's life. He had believed the world to be one way and his discovery that it was ordered differently utterly changed the meaning of both his present existence and his life up until that point. His sense of resentment at the manner in which his life had been a form of social experiment, which was abandoned before reaching its conclusion, was so great that he was able to articulate a sense of anger at the very idea of 'the journey' which the civilising ideal promoted, and to instead valorise the 'savage existence' which would otherwise have been his lot. In such irony, we also see a form of articulation of the values of indigenous cultures which I have been suggesting are writ large across his letters in subtler forms, in the manner in which French practices are set aside higher ideals whose origins lie, I think, beyond simply the standards which the colonial state set as its rationale, and can instead be seen to refer to an ethical index derived from local culture.

If we look further into ben Zahra's correspondence, we find that both his day-to-day and overarching complaints with the medical establishment in Algeria revolved around questions of justice, both as they related to him and to the community of patients to whom he owed a duty of care. The letters bluntly reveal an Algerian's view of the manner in which France had reneged on the promises of medical imperialism. They show the extent to which the French rhetorical propagandisation of that idea had infiltrated Algerian society and the gaps which had always existed between enthusiasm for the promotion of the idea and the much more testing reality of instantiating that idea in practice, given the set of constraints which existed in the colony in terms of changes of regime, administrative failures, opposition to medical imperialism, conflicts between colonial forces and the original failure to acknowledge the stark choices which would need to be made when the idea of medicine came up against leading ideas of imperial society, such as cultural racism.

Ben Zahra's sense of betrayal and injustice reflected very much a sense of this gap between the avowed beneficence of medical imperialism and the truth of the offer being made to Algerians. In 1900, we see a similar sense of moral anger in Prochaska's account of the life of Khélid Kaid Layoun:

> As part of the metropolitan French reform effort of the 1890s, now winding down, a group of senators on a fact-finding tour passed through Bône. Khélid addressed them on behalf of a group of 'young Bône Muslims'. He described the situation of these Francophone young Algerian males such as himself, who favored Algerian assimilation to France but could not convert their French education into a job, who were less and less Algerian but prevented from becoming more and more French, who were in short 'floundering in civilization'.[8]

For both ben Zahra and Layoun, this sense of 'floundering in civilization' represented a profound form of injustice that was increasingly to inform brands of political resistance to French rule, for what it alleged was that France had failed to deliver on her beneficent promise to the Algerian people. Even that tiny experimental elite to which these two men belonged had not been permitted to become Frenchmen, or to be offered a status and set of rights akin to those which had been offered to even the poorest European migrants who had become French on settling in the colony. If resistance in the first 50 years of the colony had centred around attempts at militarily repelling the invader, from the 1880s onwards the career of men such as ben Zahra shows us that it cohered around civil resistance based upon a profound sense of injustice.

Driven by a disgust with unjust systems that proffered claims to medicalised justice towards Algerians, ben Zahra described the great gap which existed between colonial medicine in theory and in practice, and the very limited will and capacity of the state to undertake systematic programmes of medicine and public health in the 1880s. In October 1884, he described a report which he had written:

> for the Commander at Biskra in which I set out the sanitary situation of the territory and a programme of hygienic and public health improvements

which were needed as a means of rapidly improving the health of the region. As with most of my letters, all I heard was silence. It was reminiscent of the manner in which I was treated when I sent a demand for pox vaccines to be sent so as to combat an imminent public health disaster as a result of the spread of an epidemic from Oued Souf. On hearing nothing, I myself, at my expense, sent a telegraph to a pharmacist in Batna to request supplies.[9]

Again we have a sense in which, even in spite of the obstacles which he faced, ben Zahra conceived of his work as a doctor as having important public health dimensions, even if the colonial medical administration saw itself as having no responsibility for such measures. Crucially, ben Zahra was interested not simply in combating disease but in actively 'improving the health of the region', yet, as we saw in the previous chapter, such proactive goals were wholly uncharacteristic of the instincts of both doctors and other colonial professionals in Algeria.

Returning to the circumstances surrounding the rejection of ben Zahra's promotion to the rank of doctor – and the relationship between questions of injustice and race – in May 1884, Senator Le Lelièxne received a reply from the Governor General to a letter he had written asking that a post as médecin de colonisation be found for his protégé, ben Zahra. The Governor General wrote that 'While he would have very much liked to have found such a post for ben Zahra, the senator needed to be aware that the regulations of 23 March 1883 allowed for officiers de santé to be appointed to such positions only when no fully-qualified doctors were available.'[10] For this reason, he was unable to help at that time. From this brief exchange we can see some of the central realities of the medical system for men like ben Zahra: they were trained as officiers de santé, rather than as doctors, to preserve a racial hierarchy in the profession; they depended on patronage to obtain any kind of medical position with the qualification they had earned; and that there was a limited demand for medical graduates of Algerian origin.

In a follow-up letter direct to the Governor General, ben Zahra pressed his case for a post as a médecin de colonisation in the Tell or, at the very least, a temporary position in Tuggurth, where two doctors had died in post in quick succession. We have seen that ben Zahra's letter set out the great difficulties he had encountered in becoming one of eight Algerian Arabs to have qualified for medical positions, and how he believed that even such slight opportunities were being denied to those of Algerian origin, in addition to career paths being closed down for this tiny elite who simply wished to serve France through medicine.

Across the files relating to ben Zahra's career, we find a dominant concern with questions of justice. In this regard, there was a rather telling contradiction in the complaints of local military authorities regarding ben Zahra's work, for at times he was castigated for his frequent absences, particularly those associated with his wife's ill health, whilst at others he was criticised for the excessive vigour he adopted in his work. There is no doubt that ben Zahra was often on leave, but such requests were always based upon either his own poor health or that of his wife. When working it is clear that he approached his task as a colonial doctor with great vigour, choosing to take responsibility for questions of public health strategy and the regional administration of healthcare services as well as his regular duties in terms of running a small hospital and travelling around the district to attend to the sick. His personnel files include a number of letters denouncing his character from military figures, who argued that the local commune would be better served by having its tax burden reduced by four thousand francs annually, rather than paying for a doctor like ben Zahra who was so rarely in residence, yet there are compelling reasons for thinking that it was as much ben Zahra's ambition as a doctor and a medical administrator which motivated such complaints from the army.

In fact one of the chief complaints of local military commanders, as seen in a letter to the Governor General of 14 April 1884, was that ben Zahra was too active in his position, for the military objected to the fact that he was claiming back money for medicine which he had dispensed free of charge to locals. This objection in fact arose from a confusion regarding ben Zahra's status in the complicated medical administration of Algeria. If he were a médecin de colonisation appointed by the central state in Algiers, and funded from the capital, then he would be entitled to act as he had, but the soldiers argued that he was actually merely an officier de santé, who ought to have been treated like other 'médecins communaux', for ben Zahra's salary, and the costs incurred from his free prescription service, were borne by local taxpayers.[11] This bind in which ben Zahra was placed was typical of the incoherence of the practice of medicine in Algeria and his medical mission was doomed to failure precisely because his role was constrained by poorly framed legislation which was expressive of political rivalries rather than medical outcomes. Again, in his commitment to the provision of free healthcare and to public health, we see intimations of the manner in which ben Zahra turned to Islamic conceptions of health and ethics as a means of describing both the injustices being done to him, to those whom he served, and to Algeria more generally.

6.5   The Tuggurth posting

Ben Zahra had eventually been appointed in Tuggurth, almost certainly because it was a posting which European doctors would not accept, no matter how ambitious they were, on the grounds of the colossal risk involved in being posted to such a place. In 1871, a French soldier who had been stationed in the town had written of the 'horrors' of its fevers and of the brutality of the culture which accompanied such a dangerous environment.[12]

Ben Zahra was well aware of these dangers, for on accepting the position in June 1884, he wrote to the Governor General to state that his desire to recognise all that the government had done for him, in taking care of his childhood and making him what he was today, deserved a practical expression of thanks.[13] This was in spite of the fact of 'the sanitary conditions of the territory there, its distance from all intellectual life, the premature deaths of Mohammed Mustapha and Djillali ben Fiah from illnesses contracted in post, and his huge desire to continue his studies so as to obtain a doctorate' (which would have seemed to have placed him on an equal footing with French doctors).

Yet the realities of the harshness of the conditions in Tuggurth were much worse than ben Zahra had envisaged. On arriving there, he had still been suffering from the after-effects of a 'fièvre paluske' which he had contracted when he had been posted at Oued Rhia. As we have read both he and his wife were sick for most of the time they lived in Tuggurth, so much so that his wife was sent to recuperate on the coast. Ben Zahra's many local enemies complained that until he was dismissed in the summer of 1885 he himself had only lived in Tuggurth for four months out of twelve, and that his loss would hardly be missed since there was also a military doctor stationed in the town.[14]

It is certainly true that ben Zahra did not adapt to the 'great heat' of Tuggurth, which is unsurprising given his own sickness.[15] Yet it was not just the temperature which he found oppressive there, for he described life in the town as consisting of 'terrible solitude, being far from family, friends and all relations with the civilised and intellectual world' to which he had become accustomed in the very different world of the coastal cities.[16] A third form of local tyranny came, as we have seen, in the manner in which he was treated by the military authorities who governed the town. Ben Zahra described a relentless campaign of intimidation against himself as a potential rival source of authority in the town, whose Algerian birth might help him to act as a more effective mediator with the locals than would be the case with the French army. The army constantly attempted to make his life difficult, preventing him from undertaking his duties, and crushed his sense of professional esteem in the most public and humiliating fashions possible.[17] He concluded his letter to the Governor General by writing that he very much wanted to make a living for himself and his wife that depended on the fruits of his labour, but that the fruit he had found in Tuggurth was unbelievably bitter.

The last we read of ben Zahra from his personnel file comes in a letter written by a Dr Picardin in December 1885.[18] The writer was a doctor based in Bouira who wrote to the Governor General to discuss the options open to them in terms of finding a replacement for a Dr Gensolles who had left his position. Picardin had recently lost both his wife and his 12-year-old daughter to diseases in Bouira, but he was determined of two things: to remain in the town and to prevent ben Zahra, who was serving as Gensolles' replacement on a temporary basis, from being appointed to a permanent position in Bouira. Picardin acknowledged that it was extremely difficult to find a doctor who was willing to serve in the town, for many had refused 'on hearing of the terrible manner in which Mayor Paoli and his councillors had treated Gensolles', but he assured the Governor General that the atmosphere in the town was now much improved.

We know no more of why Picardin objected so violently to ben Zahra, nor indeed whether he continued to serve in Bouira, though the absence of further documentation does suggest that his medical career may well have ended just years after he had graduated as an officier de santé. The suspicion must be that Picardin's fears regarding ben Zahra were similar to those which had seen him driven out of Tuggurth and which ben Zahra foresaw in his letter to the Governor General in May 1884. In the following chapter, we will see how common ben Zahra's experiences were amongst that tiny group of Algerians whom France trained as representatives of the idea of a civilising medicine. Each of these men's careers seemed to be dominated by themes of the collapse of the idea of medicine in practice and the increasing sense of anger and betrayal on the part of Algerian doctors that the civilisational promise was hollow for them as individuals and for the people from whom they came. Abdel Kader ben Zahra may not have been treated justly, but his sovereign testimony was his own record of the ethical failure of the colonial state and its medical mission.

7
On Injustice and the Disavowal of Autonomy

7.1 The life and death of Mohammed ben Mustapha

The personnel files of Mohammed ben Mustapha, who may have been the third or fourth Algerian-trained doctor to work in the colonial medical service, reveal only one significant facet of his career and one key feature of his life, which led very directly to his death, about which we know only a little more.

Like Abdel Kader ben Zahra, ben Mustapha both served in Tuggurth and was to find that his career would end there. Before that posting, he had worked in Ouazgla where, like ben Zahra, he had found that his professional practice was severely impeded by pressure from local French colonial forces, though in his case it came from Fr ench missionaries rather than the army. In 1870, he wrote to the Governor General to complain that Catholic missionaries were dispensing free drugs, 'telling people that they were the true doctors and that he was a charlatan'.[1] As ben Mustapha angrily observed, the reverse was in fact true, for while he possessed formal medical training, it was very unlikely that this was the case for the mission's 'doctors'.

This detail from ben Mustapha's career is revealing with regard to the often-complex cast of competing authorities at work in Algerian medicine, especially outside the major cities, and the absence of a strong, central authority as a means of regulating these contests. While it was true that many policy disputes went through the office of the Governor General, as we see here, that office was primarily concerned with the collection of information rather than active intervention in such disputes. And, as we saw in Chapter 5, Lavigerie and his missions operated in an aggressive fashion in the colony, determined to lay claim to territories where they felt there was any kind of power vacuum, and, drawing on the example of the original conquest of Algeria, they saw that medical imperialism could be an effective means of gaining such footholds in the interior. In 1868, it was famine which provided possibilities for missions to insinuate themselves into colonial power structures, and in 1870, the presence of a doctor of Algerian origin represented a similar instance of systemic weakness that might be exploited.

The only other significant event recorded in ben Mustapha's file was his premature death at the age of 32 in 1881. From his letters it is clear that he always understood that Tuggurth was a dirty, risky, insalubrious place, and when he fell ill, he requested a transfer out of the town for he knew that his chances of recovery would be much higher on the coast. This request was, however, refused and he was told that he could not leave until a replacement doctor had been found.[2]

His obituary in the bilingual French-Arabic official newspaper *Mobacher* is blunt in its account of the manner in which the difficulty of his work, and the dangers of the disease environment in which he toiled, contributed to his early death, but it also could not avoid claiming ben Mustapha as a heroic emblem of the idea of French medicine in Algeria. It announced that:

> He died at the young age of 32, taken by a cruel sickness which he had encountered in the course of his duties. On falling ill, ben Mustapha threw himself even deeper into his work, addressing not only his regular duties but also the underlying problems of public health in the town. It was at that moment that the efforts of work and repeated attacks of fevers started to diminish his once robust constitution. His health rapidly declined and on acquiring a chest infection he requested a transfer to the more clement coast, which would also have been closer to his family... he died murmuring one last thanks to France, the country which he had loved so sincerely.[3]

Therefore, like ben Zahra, he had seemingly been prepared to risk his life as a means of thanking the beneficent colonial state for the opportunities which it had afforded him above all his contemporaries, yet even this official obituary dared not occlude the evident contradiction in this claim. Ben Mustapha was quite clearly sacrificed in Tuggurth and there is ultimately no denial of the confiscation of his agency, for both ben Mustapha and his masters knew that 'the clement coast' would most likely have saved his life, yet instead he was forced to perform the role of the heroic tragedian who served as a complement to all those Frenchmen, most especially medical personnel, who had died in the colony, and also to the continuing glory of the idea of medical imperialism. That ben Mustapha, like his later compatriot ben Zahra, operated with broader conceptions of health and ethical duty, is made plain in the recognition that he concerned himself with 'the underlying problems of public health' as well as his 'regular duties'.

7.2   Dilali ben Fiah

As we learnt from ben Zahra, Dilali ben Fiah was the second of his predecessors to die in post at Tuggurth. His story is especially interesting for the manner in which it exposed the great gap between the idea of medicine and the realties of the rationing of funds for health, particularly after 1871.

The first we learn of ben Fiah is in April 1881, when the Director of the Medical School in Algiers wrote to the Governor General saying that he was finally in a position to supply the central government with a replacement doctor to relieve ben Fiah from his posting in Tuggurth. The tone of the letter was angry for the Director explained that the underlying reason for this delay was that in 1880 the Governor General's office had suspended funding for the annual bursary competition for indigènes, which had provided 12 bursaries for locals to attend the school since 1862.[4]

A year later, in further correspondence between the Director and the Governor General, we learn that in fact no replacement had been found for ben Fiah. This was indicative of both an abandonment of a commitment to something like an assimilated health service and also revealing of just how racially structured such a system had been at its most idealistic moment, for there were of course many doctors who could be sent to Tuggurth, but both the Director and the Governor General knew that only Algerian-born doctors would be so desperate to advance their careers that they would go there. The manner in which the Director appealed to the Governor General to recommence funding of bursaries was therefore a deeply practical one, for he described such funds as 'the means which we have for the recruitment to indigenous posts in the south'.[5]

Much of the rest of ben Fiah's file concerned the means by which his salary was paid, for as was the case with ben Zahra, there was some confusion as to what proportion of his salary was to be provided locally and what share would come from Algiers. The local General of the Division, who commanded the region, demanded further support from the office of the Governor General, for not only had the indigenous population of Biskra paid for ben Fiah through taxes, but a special sum had been raised locally to boost his income by 540 francs a year, taking it, including sums for accommodation and the maintenance of a horse, to 4210 francs annually. This extra sum had been granted in recognition of 'the exceptional situation of this doctor, who works in one of most advanced Saharan posts, where the climate is dangerously harsh at certain times of year, and where the difficulties of living and of finding suitable accommodation merited a higher wage'.[6]

The final documents in ben Fiah's file detailed plans to move him from Tuggurth to Beni Mansour, but it was evidently the case that he had contracted a fatal illness by that point, so like ben Mustapha and ben Zahra, his career was to be emblematic of the 'dangerously harsh' environment of Tuggurth.[7]

7.3 Boulouk Bachi

The career of Boulouk Bachi is revealing of a third potential source of tension for Algerian-born doctors working in the field of colonial medicine. While ben Zahra had clashed with the military authorities and ben Mustapha with missionaries, Bachi's problems arose from tensions with the local colons and their political leaders.

Our knowledge of his early career did not suggest that he was a man who sought conflict with authority, for from his qualification as an officier de santé in 1868, to his appointment in Fondouk in 1886 he appeared to have been regarded as a model employee. In 1869, he had undertaken military service, working in a military hospital at the time of the famine, while ten years later he was described as 'one of the most devoted practitioners in the whole of the colonial medical corps'.[8] An 1883 report suggested that he was a man who had good relations with his superiors, his juniors and with the central administration, while in 1885 he was promoted to the third class of the rank of officier de santé, which merited a pay increase to four thousand francs a year.[9]

When he arrived in Fondouk in 1886, he was 38 and had been practising medicine for 18 years.[10] He was married with five children and spoke Spanish, Arabic as well as French, so he was undoubtedly a great asset to the colonial medical service. Rather curiously, he was paid only 3500 francs a year at this point, which suggested that the local authorities in Fondouk had ignored the promotion which he had been awarded by the central administration.

It is evident from Bachi's file that he was aware that Fondouk was not a suitable posting for him, as he tried to resign from his post in January 1886, citing problems with the health of his family.[11] In an earlier report, the mayor of the town had praised Bachi for most of his work (noting his intelligence, morality and judgement), but the prefect had complained that his relations with the local authorities were poor and that he was neglecting his duties to such an extent that he would need to be replaced.

A set of investigations of Bachi's work then seem to have been undertaken, for in the spring of 1886 we read a letter from an official sent to review Bachi's work for the prefect. The writer concluded that 'When he is in post, I am convinced that Bachi offers all the care his clients need, and that he is in fact devoted to such work, but the problem is that he is often absent.'[12] At times, this appears to have been because Bachi was treating patients at some distance from the town (there may have been an insinuation that this was for extra income) but when the official had visited Bachi had claimed to have been in Algiers nursing his sick wife. In response to such accusations, Bachi declared that almost all of his absences were indeed caused by visits to distant farms, and that those who accused him should be well aware that he fulfilled his obligations according to the decree of March 1883 which required him to offer free consultations at set times twice a week, either in his own residence or at the town hall. He went on to add that the problem in the locality was more with the office of the mayor than himself, since the mayor had refused to supply him with a list of indigènes in the department, which was critical for the management of his service.[13]

By the end of 1886, it was clear that the level of hostility directed towards Bachi had increased for the local prefect wrote to the Governor General accusing the doctor of negligence, writing that he did not respond to demands for meetings and asserting that the mayor of Fondouk had reported that he was disregarding a number of his duties.[14] This referred to the annual report on Bachi which had been written by the mayor in which he had claimed that 'It was to be regretted that he is absent too often and the population was growing restless for they did not have confidence in his services and the fact that he was not attending to his duties.'[15]

Yet that report also contained indications that Bachi's problems with the local authorities may in fact have had little to do with questions of his absences from the town or his derogation of his duties, and much more to do with the question of whether the colons considered it appropriate for an Algerian-born doctor to become involved in questions of politics. In his note, the prefect in fact claimed that Bachi was 'a capable and active doctor', and while he did note that he should be reproached for being absent too often, it seemed that the underlying problem was that Bachi 'had become involved with questions and people which were dividing the commune'. For this reason, 'a change would be necessary.'[16]

More details of this involvement were revealed in a letter of December 1885 from the mayor of Fondouk to the Governor General, in which he complained that public officials were becoming much too involved in elections. With regard to Bachi, he alleged that 'Our médecin de colonisation is mixed up in all the elections in a very public fashion, debating and discussing such things in public places such as cafés.'[17] This leads one to ask whether the complaints directed at Bachi really referred to his absences, or to an excess of his presence in certain senses in the town. The mayor went on to say that 'on the fourth of October he was seen openly campaigning against the candidacy of Bonheur and Pelletier, and was in many senses the cause of the failure of these candidates to be elected in Fondouk'.[18]

The last we read of Bachi comes in an 1887 report which again addresses the question of his neglect of patients and his over-involvement in local politics. It is unclear whether such allegations ended his medical career or whether there was some other reason as to why his personnel file ended so suddenly.[19]

7.4 Mohammed ben Larbey

Our fifth doctor was evidently a man of great ambition, but his experiences show how difficult it had become to follow such dreams in the colonial medical service by the 1880s. Like the other cases we have looked at, ben Larbey's is a story of disappointment and an awareness of the gap between the bright idea of medical imperialism with which his career had begun and the realities of life on the edges of empire in the colonial medical service.

Ben Larbey was a great rarity for he had begun his medical studies in Paris, and on graduating as an officier de santé he had worked in Algeria for a number of years. By 1881, ben Larbey was keen to continue his studies and to return to Paris so that he might gain a doctorate and become a fully-fledged doctor. He applied, through the military authorities of the interior, to Algiers for a grant of 5000 francs, which would consist of 450 francs for a set of five exams, a similar sum for the printing of copies of his thesis, 3600 francs living expenses for two years and at least 500 francs for books, clothes and travel.[20] Ben Larbey appeared to have believed that he had secured the support of his military commanders in this aim, but the letter from the officer concerned revealed otherwise:

> In making this demand, the indigène in question invokes the favour which had already been granted in similar circumstances to a number of his fellow Muslims. Yet he is in error here, for there has in fact been only one Muslim, Ben Mekkache, who has been funded in this fashion, at a cost of 4500 to 5000 francs, to study in Paris. This example was therefore unique and the considerable costs which it entailed have convinced the central administration that such a heavy burden on the budget will not be repeated. Why should such a colossal sum be spent on the creation of indigenous pseudo-doctors, who, with their double qualifications, would lie in an awkward position *vis à vis* their French colleagues, and who would subsequently demand as of right positions in the medical service which there is no chance would become immediately available? While there might be some justification for according a few local doctors training slightly beyond their initial studies, what advantage would there be in encouraging them to obtain doctoral titles, which they might gain without having to demonstrate the skills necessary to earn them? Finally, moving beyond the question of the level of education appropriate to the indigènes, why should we offer something to Muslims which we have refused to offer to the French?[21]

The officer went on to say that it would be pointless to grant any money to ben Larbey, for that would only encourage his ambitions. He also noted that, in any case, the funds which were available for Muslim scholarships had already been allocated in 1881 for the purposes of legal training in advance of reforms to the system of Islamic justice.

Therefore, the career of ben Larbey was indicative of the distinct limits which the colonial authorities placed on the careers of Algerian-born doctors, and the diminishing importance of the promotion of the idea of medical imperialism as the century wore on. Ben Larbey's own experiences as a medical practitioner were similarly illustrative of such broader trends, for he encountered yet another form of competition as an indigenous doctor; this time from a rival French doctor who successfully lured a large proportion of his clientele away from him.

Ben Larbey worked in Bou Medja and it was clear that, like Bachi, he had considerable problems with the local class of colons from his arrival in the town. In October 1876, the local prefect wrote to the Governor General to complain that 'While his Muslim brothers go to him with all their medical needs, with trust in what they perceive to be his French methods, the reality is that his bearing shows, on the contrary, that he has no idea of how to act according to French social norms.'[22] The colons 'accepted his treatment only with repugnance' and the writer claimed that a pattern was forming, for similar complaints had been made about ben Larbey in Bourkika.[23]

Ben Larbey was evidently moved on, for when next we hear of him we find those earlier complaints replicated in the words of the prefect of Oued Fooda, who alleged that Ben Larbey 'inspired so little confidence amongst the inhabitants of the town that another officier de santé had managed to come into town and build up his own client base'.[24] The prefect contended that this situation could not be allowed to continue, for the cost of both of these doctors were being funded by the colonial medical service, and the prefect hoped that 'it might be possible to use M. Larbey in a more favourable location by sending him to a territory which was inhabited only by indigènes'.[25]

In fact, we learn that a more concrete proposal was made that ben Larbey be sent to Tuggurth, which should come as no surprise. While it is unclear as to whether the populations of these towns had any genuine reasons for feeling aggrieved at ben Larbey, the evidence would seem to suggest otherwise, for it seems more likely that his career was simply blighted by his being an Algerian Muslim sent to work with racist colons. I feel confident in saying this because there is too great a disparity between the criticisms of ben Larbey and the factual record of his career. The complaints made against him related exclusively to his failure to relate to the French and to the norms of French colonial society, yet the reality was that ben Larbey was one of only two Algerian-born doctors who had actually lived in France, and who would therefore have been completely aware of such norms. Perhaps that was indeed his true problem, for while he may have been aware of the social norms of Paris, he was perhaps less able to adapt to the 'French' norms of colon society; a culture which contained many who had had no experience of French life, let alone of a metropolitan existence.

7.5 Mohammed ben Saïah

The career of the last doctor of Algerian origin which I wish to look at is of especial interest for the extended duration of both his work and our knowledge of his career, for Mohammed ben Saïah started working in the colonial medical service before any of the other doctors mentioned and, as far as we know, his career finished later than theirs. For this reason, I propose to look chronologically at a medical career which lasted almost forty years, from the start of his medical education in 1865 to his dismissal in 1904. Ben Saïah's life also provides me with a chance to reflect upon some of the trends we have seen across the careers of the doctors studied, for although his career appeared to be a very successful one, he too eventually encountered familiar sets of problems which drove him from his work as a doctor.

The first we hear of ben Saïah is in a letter of 1865, in which a general of the second division wrote to the Governor General to encourage the government to support ben Saïah's medical education.[26] Such patronage was of course representative of one particular strand of military thinking in Algeria at that time, in which a part of the army was still very much committed to the idea of medical imperialism and the pacification of tribal areas through a process of pedagogic and medical assimilation of Algerian elites.

The next we hear of ben Saïah was in 1874 when General Carteret mentioned him with reference to his proposal for 'médecins de tribu', tribal doctors. Carteret proposed a scheme whereby Algerian doctors would be given an annual grant of 600 francs and rather than be assigned to a particular community, they would simply have had the freedom to attract a client base in the tribal areas. Such a scheme was of course to become a major source of complication in the lives of all medical professionals working in Algeria for it increased competition amongst medical providers and had the potential to destabilise even further the fragile healthcare network which notionally put into practice the tenets of the medical imperial idea.

In fact, Carteret did not have ben Saïah in mind as one his first tribal doctors because he had received information about his moral qualities which was not at all favourable.[27] Nevertheless ben Saïah was offered a position in Tuggurth later that year, with a contract that paid 3000 francs annually, with 500 francs expenses, the provision of transport, and 'authorisation to wear indigenous clothing whilst fulfilling his professional functions'.[28] Ben Saïah was to be the only Algerian doctor we know of who survived his posting in Tuggurth, for he was alive and his reputation was intact when he left the job in 1877.

In spite of the fact that his contract had promised him transportation, one of ben Saïah's major problems in Tuggurth seemed to have been securing funds to buy and maintain a horse. In October 1875, he wrote to the Governor General to request stabling facilities for a horse which he had bought, but Algiers argued that ben Saïah should rely on the local tribes rather than the central administration for such support. In special cases, it was argued, the agha could requisition a spahi horse for ben Saïah. Yet there was a problem in that the tribes lived far from the town and the spahis were extremely reluctant to loan horses, which is why ben Saïah had arrived at his own solution to this problem. The officer of the Governor General disclaimed any responsibility for such things, arguing that the commune of Biskra would have to bear the cost of ben Saïah's transportation if no other means could be found, revealing again one of the central dilemmas faced by doctors at this time, which was the considerable confusion, and ensuing conflicts, which surrounded the funding of healthcare.

This was again made apparent in ben Saïah's next posting which was to the town of Bou Saada in 1877. The community there had voted in that year to spend a proportion of their taxes on the appointment of a 'médecin indigène', but on arriving in the town ben Saïah was surprised to find that the local commune believed that it would be paying only a third of his salary; for some claimed to be shocked at the idea of paying for medical care at all, having previously been tended to for free by army doctors.[29] A series of letters went between the commune and the office of the Governor General, and between the général chef d'état major and the Directeur Général des affaires civiles et financiers, with the latter figure (representing the civil regime) insisting that the army was responsible for two thirds of the cost of ben Saïah's salary.[30] Unsurprisingly ben Saïah was g reatly concerned by these disputes and in one of his letters he referred to a plan by the commune to sell land as a means of generating funds to support his appointment.[31]

Other details of his career in Tuggurth and Bou Saada are scant, but there are hints of both tensions around his work and also official praise for his efforts. In 1878, his annual report mentions criticism of ben Saïah from two patients, one of whom was Jewish.[32] Similar reports in 1882, 1888 and 1890 seemed broadly positive, though they mentioned the fact that ben Saïah no longer made any attempt at approaching Europeans, who preferred to visit the military doctor, which fits with the general picture we have of an increase in racial stratification and zoning in towns at this time.[33] What is interesting about this state of affairs from a medical administrative point of view is that it was driven as much by patients as it was by governing authorities, for it was the decision of colons to rely on the army's medical care which was to lead to a new kind of alliance in Algeria. What I mean by this is that whereas the colons had previously identified themselves with the civil state and against the army, that in certain places a shift of allegiances was taking place, for the centralised civil state now took on more responsibility for Algerian-born doctors and resented the army's recuperation of its influence through its provision of free medical care to Europeans.

While ben Saïah's primary orientation towards the indigenous population had evidently been known about for some time, a scandal surrounding his work suddenly appeared in the documents from his personnel file in 1895, though it was clear that this referred to long-standing local tensions which had surrounded his practice. In April 1895, ben Saïah was sent on leave and in September of that year he was assigned a state of 'non activité'. He fought hard to retrieve his position, but was refused on the basis of his supposed incompetence and, seemingly more importantly, for his promotion of 'les mœurs et les coutûmes indigènes' which could not happen in 'un centre de colonisation européen'.[34] As another official letter of the time stated,

'It will be difficult to conceive of a way of using the services of M. Mohammed ben Saïah in providing medical services to Europeans, for this doctor has preserved the morals, habits and customs of the indigenous Muslims.'[35] Yet why should such complaints emerge 30 years into his medical career, having not appeared at all in his previous posting in Tuggurth and having evidently been tolerated in Bou Saada for almost twenty years? For what was 'the preservation of indigenous morals' a form of coded reference?

The answer to these questions seems to be that ben Saïah had become caught up in a set of broader political struggles in which the army stationed in the town had struggled to maintain their authority and were determined to purge the territory of all of those who they regarded as having the potential to influence the indigènes and foment trouble. The first sign we have of the problems which ben Saïah was to face came in a petition calling for his dismissal in February 1895 which was compiled by the governing body of the local commune, which included French, Jewish and Arab names.[36] Their grievance was against both ben Saïah and a cadi named Si Kaldun ben Zaieb, both of whom were described as 'arbitrary, responsible for countless iniquities and vexations, men of violence and no sense of compromise'. The signatories claimed that the two men were bullies whom the townspeople feared.[37]

I am somewhat suspicious of the origins of this particular petition and accompanying letters, for all the previous evidence seemed to suggest that while ben Saïah did indeed have poor relations, or no relations, with the colons, that he was very close to the local Muslim population. Might it have been the case that the military authorities persuaded a group of people that such a petition would be in their long-term interest, and might ben Saïah's dismissal have had more to do with economics, for the commune and army were evidently becoming annoyed that they were funding medical services that only reached the indigènes? Might it have become significant that ben Saïah was evidently a man of great local influence in a place where indigenous revolts had become more common in recent years? Such a view does seem to be supported by the remarks of the local general that 'the intrigues of ben Saïah ensured that Bou Saada was in a state of constant high tension, and since calm has accompanied his replacement by a military doctor, it would seem expedient to permanently exclude him from the town'.[38]

I would certainly not want to exclude the idea that ben Saïah was engaged in local intrigue, but my point is that the reasons given for his dismissal were essentially that he was an Algerian Muslim, and the fact that descriptions of his political activities amounted to hearsay, suggests to me that his true crime may well have been the high status which he had acquired amongst the local population.

There was in fact a dispute between the local army officers and the Governor General over whether ben Saïah would be dismissed, which refers back to the dual responsibility of governance in the système mixte. Officials in Algiers saw no grounds for dismissing the doctor,[39] while General Swiney begged the office of the Governor General to reconsider in September 1895, writing that peace has only just been re-established 'entre les principales notabilités indigènes'.[40]

Swiney certainly seemed desperate to employ almost any measures to move ben Saïah out of Bou Saada, for two months later he wrote to the commander of his division in Algiers requesting funds to fulfil ben Saïah's demand that he should be paid three months salary:

> In spite of the fact that this practitioner has, as a doctor, offered a pretty bad service to Bou Saada, where he created a great deal of trouble for me with the local authorities on account of his love of intrigue, I see no way that we can refuse this request on the grounds of his considerable family obligations.[41]

Various career solutions were proposed for ben Saïah, which included undertaking a stage at a hospital in Mustapha, which he refused.[42] Swiney argued that it was not politic to offer him another médecin de colonisation position, 'but since the Society of Missions in Africa are not far off establishing a medical service for indigenous doctors in their hospitals, we could place ben Saïah at the disposition of Père Foillard, to be used in Ghardaia'. Yet again, it is hard to escape the idea that doctors, and indigenous doctors in particular, were seen merely as units of production, easily replaced when they caused the colonial machine to malfunction more than they contributed to its smooth running, and simply transferred from one branch of the mechanism to another.

The last nine years of ben Saïah's career were evidently traumatic for where before he had established himself in positions for considerable periods of time, he now struggled to hold onto any kind of work. The options open to him were limited, for he was considered too great a risk for many kinds of medical work, most especially positions in tribal areas of hospitals where he would have the potential to influence the indigènes, yet he was also seen as being constitutionally incapable of working with and for colons.[43]

In 1897, he secured the support of an Algiers deputy, Samary, who wrote to the Governor General on his behalf, and he was given a new post in 1898, but he was suspended from that position in the same year, quite possibly as a result of a letter from general Meygret to the Governor General, reminding him of the great troubles ben Saïah had caused at Bou Saada.[44] In 1899, he led a mission to study the symptomology of skin diseases and syphilis in Kabylie, but in 1900 he again seemed to have encountered problems for a leave request of that year was refused.[45] The last we learn of him is in a letter of April 1904 in which the prefect of Algiers wrote to the Governor General explaining that ben Saïah still had a great desire for a medical posting in spite of his most recent dismissal from such a position.[46]

7.6   Philippe Grenier: a coda and conclusion

Unlike Abdel Kader ben Zahra, therefore, Mohammed ben Saïah had retained a sense of hope and faith in the systems in which he worked, in spite of his travails and the manner in which he had been treated. The colonial authorities trained only a tiny number of Algerians as medical professionals in the nineteenth century, yet their careers offer little sense that there was any practical concern with ensuring that their lives truly stood as emblems of the ideas of medical imperialism and civilisational beneficence. Instead, they were tragic figures, for their destiny was to live as remnants of an idea which had effectively died as an aspiration to practice and which would therefore seem to have waned as a means of structuring the identity of the colony.

Many of the doctors came to some sense of realisation as to the manner in which the realities of their careers were very different from the promises they had been made and those notions which they were assigned to represent. They faced common problems of insecurity, competition from a variety of other medical providers, racism, budgetary constraints, the threat of the colons and a more general lack of willingness of colonial culture to practice the assimilatory ideals which it preached – even in the case of that tiny number of individuals amongst a population of millions which it had selected as emblems of that purpose.

A number of the 'pseudo-doctors' came to see the civilisational narrative as a lie and, as I have suggested, they connect the moral critique of imperialism of men like Hamdan Khodja with the similarly ethically based excoriation of France's behaviour in Algeria which we find in the work of Fanon and late opponents of French imperialism. Like ben Zahra, they came to see their careers as being characterised by injustice and a disavowal of autonomy.

Their identification of moral failings which underpinned the colonial management of medicine was often complemented by expressions of ethical ideas about health which drew on their Arab-Islamic heritage. In some cases this was explicitly commented on by the colonial authorities, as in the case of Mohammed ben Saïah, but where the French tended to view such matters solely as proofs of failure to adapt to the norms of a more sophisticated culture, there is evidence that in their medical practices the Algerian doctors attempted to work in ways which they believed were true to their own moral heritage. This was most evident in relation to questions of public health, which featured in the careers of a number of doctors, in spite of the frequent opposition of French authorities, who tended to view such work as moving dangerously across administrative and professional borders. Yet this was evidently the point for men like ben Mustapha and ben Zahra for they well understood that real improvements to the lives of their patients could only be achieved through the coordinated planning of health which refused to see individuals and settlements as isolated outposts with no connection to their broader environment. The sense of injustice which characterised many of their careers and the attacks which they felt on their autonomy were also signs that rather than serving as symbols of a beneficent imperialism, they stood as proofs of the complexity of the ethical encounter which occurred in the medical sphere in Algeria.

As a means of concluding this book, I wish to explore the career of another doctor working in France and Algeria at this time as a form of coda to my descriptions of the careers of Algerian-born doctors working in the colonial medical system. I hope that a description of the life of Philippe Grenier serves as a form of contrast and counterpoint with the realities of the lives of the Algerian doctors, revealing new aspects of the manner in which questions of identity and the relationships between France, Algeria and Islam functioned at the end of the nineteenth century. We shall also see that Grenier's career revealed the curious longevity of the power of the idea of medical imperialism in spite of both France's evident failure to medicalise Algerian society in the nineteenth century and, relatedly, the subjugation of local cultures and the deterioration of local bodies at this time.

Grenier was a complex figure whose lack of renown belies the importance he might be accorded in a history of modern France. He began his career in 1890 as a rather unremarkable medical doctor from the Alpine town of Pontarlier in the Daubes, but his life was to take a series of interesting set of turns after he travelled to Algeria. His father had fought in Algeria earlier in the century and Grenier went there to visit his brother who had continued the family tradition of military service. What he saw there had a profound effect on him for he was struck by 'the poverty of the local population, by their abandonment and their resignation'.[47] Yet alongside this dismay at the failure of the French to accord Algerians lives of dignity, he also developed a profound respect for Islam, and on a series of trips to the country he spent a great deal of time trying to learn as much as he could about the religion. In 1894, at Blida, he converted to Islam and decided to both undertake a pilgrimage to Mecca and to adopt a traditional mode of Algerian dress, which he wore for the rest of his life.[48]

When this radicalised and newly exotic figure returned to France, he was questioned as to why he had become a Muslim and what advantages Islam possessed, to which he answered:

> You wish to know why I became a Muslim? The answer is a combination of taste, character and belief, but it has nothing to do with a form of fantasy, as some have insinuated. From a young age, Islam and its doctrine have held a powerful attraction for me. In the course of a series of trips to Algeria, what had initially been an interest became a kind of fervour, but a reasoned fervour, for it was through a close reading of the Qur'an and deep study and meditation of the faith that I embraced Islam. I adopted this faith and its dogma because it seemed rational to me and much more

closely allied to science than is Catholic faith and dogma. I should add that the prescriptions of Islamic law seemed to me excellent because, from a social perspective: Arab society is based on the family and principles of equity, justice and charity towards the weak. What is more, from the point of view of hygiene – which is evidently of great importance to a doctor – Islam proscribes the consumption of alcohol and carefully orders ablutions and the washing of clothes.[49]

This answer is fascinating in a whole series of ways: as an implied critique of the lack of ethics which underpinned French rule in Algeria, in his critique of the idea of Algeria and Islam as a form of 'fantasy', of the powerful persistence of Islamic morals and ways of life in the colony, in his perception of Islam as a religion of science, and the manner in which that is an example of his more or less complete rejection of the colony as it stood, as well as the fact that his views were wholly opposed to those of men like Bertherand, Bodichon and the leading colonial doctors. In looking at Grenier's life, it initially appears as though we are dealing with a form of diametrically opposed 'fantasy' to that enacted by the French through medicine in Algeria, for where the colonial power attempted to make Algerians French through this fantastic idea, Grenier seemed to be suggesting that it was his European medical-scientific background which had led him to comprehend the virtues of adopting an Algerian Muslim identity. In this respect, he would seem to be a great rarity.

Grenier's travels and his reading had left him very familiar with the French critique of Islam and the manner in which this had been deployed in the moral conquest of Algeria, for he saw this interpretation as being immoral in the manner in which its consequences had impoverished the lives of Algerians. He rejected, for instance, the French obsession with fatalism as a sign of primitivism, for as Grenier wrote, fatalism was a very specific notion which referred only to 'Resignation to the will of God when it has been enacted, and when all other options have disappeared, which did not prevent the idea of struggle for those things which were still to come.'[50]

He was also attentive to the great gaps which had emerged between French rhetoric and the practice of rule, asking why 'when Crémieux had naturalised the Jews of Algeria, had Muslims not been naturalised too? Had they not merited it given the fact that Algerian Muslims had served and died in French armies for the flag and the honour of France?'[51] In stressing the loyalty of Algerian Muslims to France, Grenier was at one and the same time deploying an argument which was likely to be accorded respect in France and, I think, revealing the manner in which his Islamic identity did not originate quite as much in the deserts of Algeria as his mode of dress seemed to suggest.

Grenier would have been of interest simply as a French doctor who had converted to Islam, who had seen his medical science and new faith as being

highly compatible and who had criticised the practise of French imperialism on the basis of the shared ethics which he had found in science and
Islam. Yet perhaps the most remarkable fact of his life was that he was
elected as deputy for Pontarlier in 1896: the first Muslim to sit in either the
Chamber of Deputies or the Senate (indeed, he is the only metropolitan
Muslim to have ever sat in the Chamber). The town was a Radical, Catholic
stronghold, so Grenier's victory, standing as an independent, was seen as
being all the more remarkable by the national press.

In the first round of voting Grenier had actually come third, but had
refused to withdraw from the race, saying 'I have no right to do so ... for
Allah wishes that I should teach his religion in the Chamber.'[52] In the second round, Grenier benefited from the great unpopularity of the incumbent
deputy, Grillet, and was elected primarily through his acquisition of a large
number of second-choice votes. He was ridiculed by much of the local press
in the course of election, with a local alpine paper, the *Courrier de la Montagne*,
mockingly announcing that 'when it comes to the Prophet of God, he thinks
he will be able to realise the promise made by Mohammed: to bring him to
the mountain-top'.[53]

His candidacy was also used by others as a means of talking more broadly
about the political situation of France and of Algeria in the 1890s. The anti-
Semitic paper *Le Soleil* declared that 'when there are hundred deputies in the
French chamber representing the tribe of Israel, I think it reasonably just
that there should be one who serves to defend the rights of the four million
Arabs and Kabyles that the *racaille juive* exploits, holds to ransom and
oppresses'.[54] Similarly, Jean Jaurès saw a socialist lesson in the election of
Grenier, announcing that the 'Algerian Arab' was

> [s]uspected by the colon, ruined by the Jew, systematically deprived of his
> land by statute and procedure, kept out of higher education, his civilisa-
> tion and his home ruined, and ignored and disdained in the metropole,
> living in a sad ignorance, a hopeless and hateful misery. But at the same
> time, the force of their thought, the beauty of their poetry, the power of
> their piety and the rhythm of the Qur'an had seduced a Frenchman from
> France, who now prepared to enter parliament to defend the Arab people
> with a wholly new authority, which had been granted to him by an entire
> community of faith and of beaten souls.[55]

Leaving aside the task of unpicking the Orientalism and anti-Semitism of
Jaurès' speech, what seems to me quite remarkable is how French commentators took Grenier's claims to be the deputy of the Arabs at face value. In dress-
ing as an Algerian and describing himself as a Muslim it was as though he had
actually become these things, and while it is certainly true that Grenier was
an impassioned defender of Algerians, it is equally notable that he had first
travelled to Algeria, converted to Islam and been elected as a deputy in the

space of a few years. His knowledge of both Islam and, more particularly, the complexities of Algerian society, was profoundly limited, as was his understanding of the structures of French power in the colony. I do not wish to diminish his uniqueness but it must be admitted that this humanitarian representative of the Algerians achieved nothing in his time as a deputy, either in terms of introducing specific legislation or in contributing to a change in parliamentary culture and debates regarding Algeria. Grenier had an underlying faith in French imperialism and, if anything, he resembled much more the spirit of moral imperialism of the 1830s than he did the Islamic resistance to French rule of the 1930s.

We should also note that there was a touch of hyperbole about his uniqueness, for while he described himself as the deputy of the Arabs, scores of French deputies had served as representatives of Algerian constituencies through the nineteenth century. Most may not have conceived of their responsibilities to their non-voting Arab subjects as their chief duty, but there had been other progressives in the mould of Grenier who had advocated the establishment of a liberal, progressive colony, most notably in the *arabophile* phase of Napoleon III's empire.

Grenier served in the chamber for only 15 months, later returning to medical practice in Pontarlier, so it seems unjust to tax him with failing to institute real change with regard to Algeria. There was after all no doubt that he was perceived as having neglected the interests of the constituents in Pontarlier who had elected him in favour of his role as a representative for Algerians.[56] His standing and his actions in the chamber saw him viewed as an exemplar of social justice, and the fact that he was French encouraged commentators to make connections between his previous practice as a doctor, where he had not charged poor patients for his services, and his stress on the connection of these ideas to Islam.

The proposals which he put forward for France and for Algeria were indeed intimately connected. For France he put forward a public health programme, which included a plan for free baths in all French communes and a scheme for combating alcoholism: in other words, this was an attempt to subtly import what he saw to be some of the best and most progressive aspects of Islamic culture.[57] For Algeria he proposed initiatives in areas such as public health, agriculture and law (where French expertise lay), and though much praised by other deputies, these plans were never put forward as legislation.[58]

Grenier offered something of a critique of French imperialism in Algeria as it had been hitherto practised. For instance, he contended that it was false that 'The Arab race had fallen into a state of decadence', though he somewhat undermined this assertion by observing that 'it too had had its time of grandeur and a civilisation comparable in greatness to that of France'.[59] He went on to reveal himself to be a thinker very much in the tradition of the moral conquest of the 1830s and 1840s, for in praising education, he wrote that it was the only means to 'reorient this people onto the path of progress and civilization'.[60] Medicine was also to play an important role in this process, for a new class of doctors was to be sent out to tend to the Arabs, distributing medicines and instructions for their use in Arabic.[61]

A similarly bipartite critique of the colony, which eventually revealed itself to be not quite as radical as it first appeared, came in his acknowledgement that 'We have committed more than one mistake in Algeria since the Conquest, and the greatest of those was the manner in which we distributed available land ... for in doing so we have created an indigenous proletariat.'[62] Yet Grenier's alarm at this situation was not focused on the great suffering of the Algerian dispossessed, but on 'the dangers which this army of the unfortunate might one day pose to the security of the colony', which elicited cries in the chamber of 'Très bien! Très bien!'[63]

In becoming a Muslim, and what he believed to be something of a hybrid Franco-Algerian, Grenier could evidently not conceive of a political situation for Algeria that was anything other than a colonial state, for his very identity spoke of the advantages of this for both parties. Grenier believed in ideas of freedom for Algerians but such liberty was to be found through her relationship with France. His underlying politics was very much in the nineteenth-century French synthetic or Saint-Simonian tradition, whereby he may have at times sounded like a socialist, but he was not prepared to make the kind of hard choices and enemies which that would have entailed. He was willing to deploy socialist ideas of 'concord, mutual benevolence, eternal friendship and fraternity between all those whose hearts were truly French', but he disdained 'the fratricidal hatred and social struggle' which others on the left had concluded would be the only means to arriving at the social goods he valorised.[64] Grenier was not alone in this naïveté, but I think he is deserving of special censure for in many ways the nineteenth century had been a disaster for Algerians, yet to read Grenier is to believe that the realisation of better times was imminent because of France's good intentions, with no acknowledgement that a politics and a conquest predicated on beneficence might conceal other outcomes.

For Grenier the ultimate purpose of Algeria was to serve France, and the greatest gift which she could offer to the metropole was a means of addressing the numerical disadvantage which France faced with regard to Germany. Post-1871, the colony therefore had a new medical-demographic purpose, for it could provide 'experienced fighters, intrepid and excellent horsemen', who might be trained in new colonial military academies. What is more, the officers trained there could return to their tribes to 'work against fanaticism' by preaching the virtues of 'nation and humanity' to their compatriots.[65] The great irony of Grenier's suggestion, which revealed his shallow knowledge of the colony, was that its praise of Algerian natality took no account of the fact that as he wrote the Algerian population was only then beginning to recover to the levels of 1830, and that death rates for Muslims were of an order well beyond that of other groups in Algeria. When we set the work of Grenier aside that of the realities of the careers of the 'pseudo-doctors', we realise that his was essentially a new manifestation of a fantasy of medical imperialism which had been implanted from the very earliest days of the colony. His innovative qualities allowed both himself and French publics to believe once more in the old lie that France had brought gifts of medicine and health to the Algerian people.

Notes

1 Introduction

1. André Nouschi, *Enquête sur le niveau de vie des populations rurales constantinois de la conquête jusqu'en 1919* (Paris: PUF, 1961), p. ix.
2. Nouschi, *Enquête sur le niveau de vie des populations rurales constantinois*, p. ix. Finding suitable terms to describe locals and invaders in nineteenth-century Algeria is a problematic business. I have chosen to refer to these communities as 'indigènes' and 'colons' on the basis that these were the designations most often used in the nineteenth-century French texts and documents from which I develop my argument.
3. Nouschi, *Enquête sur le niveau de vie des populations rurales constantinois*, p. ix.
4. Nouschi, *Enquête sur le niveau de vie des populations rurales constantinois*, p. ix.
5. E. Pellissier de Raynaud, *Lettre à M. Desjobert sur la question d'Alger* (Algiers: Imprimerie du Gouvernement, 1837), pp. 1–2.
6. Frantz Fanon, *Towards the African Revolution* (London: Monthly Review Press, 1967), p. 53.
7. Mimmi Mortimer (ed.), *Maghrebian Mosaic* (Boulder: Lynne Reiner, 2001), p. 20.
8. Alastair Campbell, Grant Gillett, and Gareth Jones, *Medical Ethics* (Oxford: Oxford University Press, 1997), p. 41.
9. Review of *Bewell, Romanticism and Colonial Disease*, p. 145.
10. Gustave Guillaumet, *Catalogue des tableaux, dessins, pastels et aquarelles provenant de l'atelier G. Guillaumet* (Paris: Imprimerie de l'art, 1888), p. 39; Gustave Guillaumet, *Tableaux algériens* (Paris: Plon, 1888), pp. 96, 198.
11. Guillaumet, *Tableaux algériens*, pp. 233–34.
12. Guillaumet, *Tableaux algériens*, p. 236.
13. Guillaumet, *Tableaux algériens*, p. 236.
14. Patricia M. Lorcin, *Imperial Identities: Stereotyping, Prejudice and Race in Colonial Algeria* (London: I.B. Tauris, 1995), p. 120.
15. Lorcin, *Imperial Identities*, p. 120.
16. Charles-Robert Ageron, *Politiques Coloniales au Maghreb* (Paris: PUF, 1972), p. 79.
17. Alan Bewell, *Romanticism and Colonial Disease* (Baltimore: The Johns Hopkins University Press, 1999), p. 1.
18. Andrew Cunningham and Bridie Andrews (eds), *Western Medicine as Contested Knowledge* (Manchester: Manchester University Press, 1997), p. vii.
19. Michael Worboys, *Reviews, Social History of Medicine*, 8–2 (1995), p. 334.
20. Jill Dias, 'Famine and Disease in the History of Angola, c.1830–1930', *Journal of African History*, 22 (1981), pp. 349–78.
21. James E. Mclellan III, 'Science, Medicine and French Colonialism in Old-Regime Haiti', in Teresa Meade and Mark Walker (eds), *Science, Medicine and Cultural Imperialism* (London: Palgrave Macmillan, 1991), pp. 36–59.
22. Mclellan III, 'Science, Medicine and French Colonialism', p. 36.
23. Richard H. Grove, *Green Imperialism: Colonial Expansion, Tropical Island Edens and the Origins of Environmentalism, 1600–1860* (Cambridge: Cambridge University Press, 1995).

24. Mclellan III, 'Science, Medicine and French Colonialism', p. 54.
25. Mclellan III, 'Science, Medicine and French Colonialism', p. 47.
26. Mclellan III, 'Science, Medicine and French Colonialism', p. 49.
27. Roger Cooter, Mark Harrison and Steve Sturdy (eds), *War, Medicine and Modernity* (Stroud: Sutton, 1998), p. 11.
28. Anne La Berge and Mordechai Feingold (eds), *French Medical Culture in the Nineteenth Century* (Amsterdam: Rodopi, 1994), p. 2.
29. Jacques Léonard, *Les Medecins de l'ouest au XIXe siècle, 3 vols* (Paris: Honoré Champion, 1978), p. 1278.
30. E.L. Bertherand, *Médecine et hygiène des Arabes* (Paris: Germer Baillière, 1855), p. 10.
31. Bertherand, *Médecine et hygiène*, p. 10.
32. Bertherand, *Médecine et hygiène*, p. 11.
33. Bertherand, *Médecine et hygiène*, pp. 11–12.
34. Bertherand, *Médecine et hygiène*, p. 15.
35. Bertherand, *Médecine et hygiène*, p. 47.
36. E.L. Bertherand, *Hygiène du colon en Algérie* (Paris: Challamel Aîné, 1875), pp. 6–7.
37. Bertherand, *Hygiène du colon*, p. 7.
38. Bertherand, *Hygiène du colon*, p. 8.
39. Bertherand, *Hygiène du colon*, p. 8.
40. E.L. Bertherand, *Hygiène musulmane*, 2nd edn (Paris: Challamel, 1874), p. 3.
41. Bertherand, *Hygiène musulmane*, pp. 3–4.
42. Shula Marks, 'What is Colonial about Colonial Medicine? And What has Happened to Imperialism and Health?' *Social History of Medicine*, 10–2 (1997) pp. 205–19, p. 215.
43. Robert Baker, 'The History of Medical Ethics', in W.F. Bynum and Roy Porter (eds), *Encyclopedia of the History of Medicine*, 2 vols (London: Routledge, 1993), vol. I, p. 852.
44. Roger Cooter, 'The Resistible Rise of Medical Ethics', *Social History of Medicine*, 8–2 (1995), pp. 257–70, p. 261.
45. Tom L. Beauchamp and James F. Childress, *Principles of Biomedical Ethics*, 4th edn (Oxford: Oxford University Press, 2001), p. 3.
46. Stephen Toulmin, 'How Medicine Saved the Life of Ethics', *Perspectives on Biology and Medicine*, 25 (1980–81), pp. 736–50.
47. Stanley Joel Reiser, Arthur J. Dick and William J. Curran (eds), *Ethics in Medicine: Historical Perspectives and Contemporary Concerns* (Cambridge, Mass.: The MIT Press, 1977).
48. Paul U. Unschuld, *Medical Ethics in Imperial China: A Study in Historical Anthropology* (Berkeley: University of California Press, 1979); M.B. McIlrath, 'A History of Medical Ethics in the Non-Christian World Before the Rise of Modern Medicine', 1959, thesis, University of Sydney.
49. F. Rahman, *Health and Medicine in the Islamic Tradition* (New York: Crossroad, 1987), pp. 91–109; Vardit Rispler-Chaim, *Islamic Medical Ethics in the Twentieth Century* (Leiden: E.J. Brill, 1993); and Azim A. Nanji, 'Medical Ethics and the Islamic Tradition', *Journal of Medicine and Philosophy*, 13 (1988), pp. 257–75.
50. Alastair Campbell, ' "My Country Tis of Thee": The Myopia of American Bioethics', *Medicine, Health Care and Philosophy*, 3 (2000), pp. 195–98.
51. Cooter, 'The Resistible Rise', p. 270.
52. Rahman, *Health and Medicine*, pp. 93–99.

53. Frantz Fanon, *L'An V de la revolution algérienne*, 2nd edn (Paris: François Maspéro, 1960), pp. 112–13.
54. See David Carr, Thomas R. Flynn and Rudolf A. Makkreel (eds), *The Ethics of History* (Evanston: Northwestern University Press, 2004); Edith Wyschogrod, *An Ethics of Remembering: History, Heterology and the Nameless Others* (Chicago: University of Chicago Press, 1998).
55. Tzvetan Todorov, *The Conquest of America: The Question of the Other* (Oklahoma: Oklahoma University Press, 1993); William Gallois, 'Todorov's Gift of Ethics to History', *Canadian Review of Comparative Literature*, 31–2 (2004), pp. 195–210.
56. Jean-François Malherbe, 'Orientations and Tendencies of Bioethics in the French-speaking World', in Roberto dell'Oro and Corrado Viafora (eds), *Bioethics: A History – International Perspectives* (San Francisco: ISP, 1996), pp. 120–21.
57. Patrick Weil, 'Le statut de musulmans en Algérie, une nationalité française dénaturée' in Françoise Banat-Berger (ed.), *La Justice en Algérie 1830–1962* (Paris: La documentation Française, 2005), pp. 95–109, p. 109.
58. Such a view also fits Ageron's general thesis regarding the manner in which the construction of the French empire in Algeria induced its own collapse.

2 On the Idea of Medical Imperialism

1. *Réfutation de l'ouvrage de Sidy Hamdan ben Othman Khoja intitulé Aperçu historique et statistique sur la régence d'Alger* (Paris: Éverat, 1834), p. 59.
2. *Réfutation de l'ouvrage de Sidy Hamdan*, p. 40.
3. *Réfutation de l'ouvrage de Sidy Hamdan*, p. 41.
4. Abdeljelil Temimi, *Recherches et Documents d'Histoire Maghrébine: L'Algérie, la Tunisie et la Tripolitaine (1816–1871)* (Tunis: Revue d'Histoire Maghrébine, 1980).
5. Temimi, *Recherches et Documents*, pp. 23, 26.
6. Temimi, *Recherches et Documents*, p. 26.
7. Temimi, *Recherches et Documents*, pp. 22–23.
8. Temimi, *Recherches et Documents*, pp. 22–23.
9. Temimi, *Recherches et Documents*, pp. 22–23.
10. *Réfutation de l'ouvrage de Sidy Hamdan*, p. 26.
11. *Réfutation de l'ouvrage de Sidy Hamdan*, p. 30.
12. Le Capitaine Villot, *Mœurs, coutûmes et institutions des indigènes de l'Algérie* (Constantine: L. Arnolet, 1871), p. 155.
13. Nacereddine Saidouni, *L'Algerois Rural a la fin de l'époque Ottomane (1791–1830)* (Beirut: Dar al-gharb al-Islami, 2001), p. 267.
14. Saidouni, *L'Algerois Rural*, p. 365.
15. Saidouni, *L'Algerois Rural*, p. 359.
16. Saidouni, *L'Algerois Rural*, p. 359.
17. Saidouni, *L'Algerois Rural*, p. 362.
18. M. Ullmann, *Islamic Medicine* (Edinburgh: Edinburgh University Press, 1978), p. 17.
19. Ullmann, *Islamic Medicine*, p. 35.
20. Ullmann, *Islamic Medicine*, p. 23.
21. Ullmann, *Islamic Medicine*, p. 24.
22. Ullmann, *Islamic Medicine*, p. 40.
23. Julia Clancy-Smith, 'The Shaykh and his Daughter: Coping in Colonial Algeria', in Burke III, Edmund and David N. Yaghoubian (eds), *Struggle and Survival in the*

Modern Middle East, 2nd edn (Berkeley: University of California Press, 2006), pp. 119–36.

24. Saidouni, *L'Algerois Rural*, pp. 359–60.
25. Docteur A. Marty, *Islamisme: Mœurs médicales et privées – climatologie de l'Algérie – considérations sur l'atmosphère* (Monaco: Petit Monagesque, 1903), p. 11.
26. Yvonne Turin, *Affrontements culturels dans l'Algérie coloniale: écoles, médecines, religion, 1830–1880* (Paris: François Maspéro, 1971), p. 89.
27. M. Rozet, *Voyage dans la régence d'Alger ou description du pays occupé par l'armée française en Afrique*, 3 vols (Paris: Arthus Bertrand, 1833), vol. II, p. 304.
28. Rozet, *Voyage dans la régence d'Alger*, II, p. 311.
29. Rozet, *Voyage dans la régence d'Alger*, II, p. 306.
30. Rozet, *Voyage dans la régence d'Alger*, II, p. 311.
31. Rozet, *Voyage dans la régence d'Alger*, II, p. 312.
32. Rozet, *Voyage dans la régence d'Alger*, II, p. 312.
33. Rozet, *Voyage dans la régence d'Alger*, II, p. 313.
34. Rozet, *Voyage dans la régence d'Alger*, II, p. 317.
35. Rozet, *Voyage dans la régence d'Alger*, II, p. 317.
36. Lucien Leclerc, *Histoire de la Médecine Arabe – Exposé complet des traductions du grec – Les sciences en orient – Leur transmission à l'occident par leur traductions latines*, 2 vols (Paris: Ernest Leroux, 1876), pp. 3–5.
37. J.A. Battandier and Louis Trabut, *Algérie: Plantes Médicales: Essences et Parfums* (Algiers: Giralt, 1889), pp. 5–6.
38. Eugène Guernier (ed.), *Encyclopédie de l'empire français: Algérie et Sahara* (Paris: Encyclopédie de l'empire français, 1946), article by Dr Grenouilleau on history of health, p. 185.
39. Guernier, *Encyclopédie de l'empire français*, p. 185.
40. A. Ledentu, *Pourquoi l'Algérie a-t-elle jusqu'ici un fardeau pour la France?* (Paris: G.A. Dentu, 1845), p. 5.
41. D.G. Trapani, *Alger tel qu'il est ou tableau statistique, moral et politique de cette régence* (Paris: L. Fayolle, 1830), p. 1.
42. Trapani, *Alger tel qu'il est*, p. 2.
43. A. Bertherand, 'De la prostitution en Algérie', in A.J.B. Parent-Duchatelet (ed.), *La Prostitution de Paris, tome 2* (Paris: J.B. Baillière, 1857), p. 536.
44. Trapani, *Alger tel qu'il est*, p. 102.
45. Eugène Bodichon, *Hygiène à suivre en Algérie: Hygiène Morale* (Algiers: Rey, Delavigne et Compagnie, 1851), p. 19.
46. Marie-Noëlle Bourguet, Bernard Lepetit, Daniel Nordman and Maroula Sinarellis (eds), *L'Invention scientifique de la Méditerranée: Egypte, Morée, Algérie* (Paris: Editions EHESS, 1998), p. 23.
47. Bourget, *L'Invention scientifique*, p. 27.
48. Bourget, *L'Invention scientifique*, p. 24.
49. Anon, *Les Princes en Afrique* (Paris: J. Delahaye, 1845), p. i.
50. *Les Princes en Afrique*, p. i.
51. *Les Princes en Afrique*, p. i.
52. *Les Princes en Afrique*, p. i.
53. Hester Burton, *Barbara Bodichon 1827–1891* (London: John Murray, 1949), p. 89.
54. M. le Baron Vialar, *Simples faits exposés à la Réunion Algérienne du 14 avril 1835* (Paris: Firmin Didot), p. 5.
55. Turin, *Affrontements culturels*, p. 14.
56. Lorcin, *Imperial Identities*, pp. 297–98.

57. Bernard Lepetit, 'Missions scientifiques et expéditions militaires: remarques sur leurs modalités d'articulation' in Marie-Noëlle Bourguet, Bernard Lepetit, Daniel Nordman and Maroula Sinarellis (eds), *L'Invention scientifique de la Mediterranée: Egypte, Morée, Algérie* (Paris: Editions EHESS, 1998), 97–116, p. 101.
58. Lorcin, *Imperial Identities*, p. 118.
59. Hélène Gill, *The Language of French Orientalist Painting* (Lewiston: Edwin Mellen Press, 2003), p. 59.
60. Deborah Cherry, *Beyond the Frame: Feminism and Visual Culture, Britain 1850–1900* (London: Routledge, 2000), pp. 79–80.
61. Cherry, *Beyond the Frame*, p. 80.
62. Cherry, *Beyond the Frame*, p. 98.
63. Adolphe Salva, *Quelques considérations hygiéniques sur Alger et ses habitans*, doctoral thesis, *Faculty of Medicine at Montpellier* (Montpellier: Auguste Ricard, 1832), p. 3.
64. Turin, *Affrontements culturels*, p. 19.
65. Turin, *Affrontements culturels*, p. 150.
66. Lorcin, *Imperial Identities*, p. 120.
67. Turin, *Affrontements culturels*, p. 85.
68. Turin, *Affrontements culturels*, p. 85.
69. Le Maréchal-de-camp B. Létang, *Des moyens d'assurer la domination française en Algérie* (Paris: Anselin, 1840), pp. 48–49.
70. Létang, *Des moyens d'assurer la domination française*, pp. 48–49.
71. Jacques L. Kob, *L'Algérie: un moyen pratique pour faire un pas en avant* (Paris: Sandoz and Fischbacher, 1880), p. 3.
72. M. le lieutenant-général Le Pays de Bourjolly, *Considérations sur l'Algérie, ou les faits opposés aux théories* (Paris: Chez Tresse, 1846), p. 1.
73. Eugène Renault, *Première lettre à M. Passy, député, rapporteur du budget du ministre de la guerre, pour l'année 1836* (Paris: Panseron-Pinard, 1835), p. 3.
74. Lepetit, 'Missions scientifiques', p. 100.
75. Lepetit, 'Missions scientifiques', p. 101.
76. Lapeyssonnie, *La Médecine Coloniale: Mythes et Réalités* (Paris: Seghers, 1988), pp. 14–15.
77. Lapeyssonnie, *La Médecine Coloniale*, p. 258.
78. Volland, M. Le baron, *Réfutation du rapport de la commission du budget, en ce qui concerne nos possessions en Afrique* (Paris: L.E. Herhan, 1835), p. 18.
79. Gaétan Citati, *Essai sur la nécessité de créer une vice-royauté en Algérie* (Marseille: Carnaud, 1847), p. 7.
80. Citati, *Essai sur la nécessité*, p. 8.
81. Le Pays de Bourjolly, *Considérations sur l'Algérie*, p. 47.
82. Charles-Robert Ageron, *Le gouvernement du général Berthezène à Alger en 1831* (Paris: Editions Bouchene, 2005), p. 165.
83. Renault, *Première lettre à M. Passy*, pp. 16, 18.
84. Ageron, *Le gouvernement du général Berthezène*, pp. 163–67.
85. *Tableau de tous les traitements et salaires payés par l'état d'après le budget de 1830*, par un membre de la société de statistique de France (Paris: Hautecoeur-Martinet, 1831), p. 21.
86. *Tableau de tous les traitements*, p. 5.
87. *Tableau de tous les traitements*, p. 7.
88. Léon Béquet, and Marcel Simon, *Répertoire du droit administratif: Algérie – gouvernement, administration, législation*, 3 vols (Paris: Paul Dupont, 1883), I, p. 228.
89. Béquet, *Répertoire du droit administratif*, II, p. 9.

90. Béquet, *Répertoire du droit administratif*, II, p. 9.
91. Maxime Laignel-Lavastine and Raymond Molinéry, *French Medicine* (New York: Paul B. Hoeber, 1934), p. 153.
92. Bertrand Taithe, *Defeated Flesh: Welfare, Warfare and the Making of Modern France* (Manchester: Manchester University Press, 1999), p. 175.
93. Taithe, *Defeated Flesh*, p. 175.
94. Georges Fleury, *Comment l'Algérie devint française 1830–1848* (Paris: Perrin, 2004), p. 244.
95. Baron Baude, *L'Algérie* (Paris: Arthus Bertrand, 1841), p. i.
96. Baude, *L'Algérie*, pp. 372–73.
97. Baude, *L'Algérie*, p. 374.
98. Baude, *L'Algérie*, p. 377.
99. Baude, *L'Algérie*, p. 381.
100. Turin, *Affrontements culturels*, p. 77.
101. Turin, *Affrontements culturels*, p. 78.
102. Turin, *Affrontements culturels*, pp. 13–14.
103. Bertherand, *Médecine et hygiène*, p. 557.
104. Bertherand, *Médecine et hygiène*, p. 558.
105. M. Bugeaud, De l'établissement de légions de colons militaires dans les possessions françaises du nord de l'afrique (Paris: Firmin Didot, 1838), pp. 1–2.
106. Bugeaud, De l'établissement, pp. 5–6.
107. Bugeaud, De l'établissement, p. 6.
108. Bugeaud, De l'établissement, pp. 7–8.
109. Bertherand, De l'établissement, p. 556 and Turin, *Affrontements culturels*, p. 13.
110. Turin, *Affrontements culturels*, pp. 81–82.
111. Turin, *Affrontements culturels*, p. 79.
112. Turin, *Affrontements culturels*, pp. 96–97.
113. Béquet, *Répertoire du droit administratif*, II, pp. 35–36.
114. Béquet, *Répertoire du droit administratif*, II, p. 36.
115. Kenneth J. Perkins, *Qaids, Captains and Colons: French Military Administration in the Colonial Maghrib 1844–1934* (New York: Africana, 1981), pp. 133–34.
116. Perkins, *Qaids, Captains and Colons*, p. 133.
117. Bertherand, *Médecine et hygiène*, p. 563.
118. Bertherand, *Médecine et hygiène*, p. 564.
119. Bertherand, *Médecine et hygiène*, pp. 558–59.
120. Bertherand, *Médecine et hygiène*, p. 559.
121. Bertherand, *Médecine et hygiène*, p. 560–61.
122. Bertherand, *Médecine et hygiène*, p. 561–62.
123. LaBerge and Feingold, *French Medical Culture*, pp. 2–3.
124. E. Delpech de Saint-Guilhem, *Mémoire au roi et aux chambres par les colons de l'Algérie* (Paris: De Rignoux, 1847), p. 6.
125. M. le lieutenant general le Pays de Bourjolly, *Projets sur l'Algérie* (Paris: Librairie Militaire de J. Dumaine, 1847), p. 24.
126. A. Warnery, *Résumé de la situation morale et matérielle de l'algérie* (Paris: E. Brière, 1847), p. 14.
127. Warnery, *Résumé de la situation*, p. 22.
128. Jules Duval and Dr Auguste Warnier, *Bureaux Arabes et Colons: réponse au constitutionnel pour faire suite aux lettres de M. Rouher* (Paris: Challamel Ainé, 1869), p. 113.

3   On Humanitarian Desire

1. Pam Hirsch, *Barbara Leigh Smith Bodichon 1827–1891: Feminist, Artist and Rebel* (London: Chatto & Windus, 1998), p. 89.
2. Burton, *Barbara Bodichon*, pp. 81–82.
3. Burton, *Barbara Bodichon*, p. 87.
4. Burton, *Barbara Bodichon*, pp. 87–88.
5. Hirsch, *Barbara Leigh Smith Bodichon*, p. 120.
6. Burton, *Barbara Bodichon*, p. 83.
7. Hirsch, *Barbara Leigh Smith Bodichon*, p. 138.
8. Burton, *Barbara Bodichon*, p. 89.
9. Burton, *Barbara Bodichon*, p. 93.
10. Hirsch, *Barbara Leigh Smith Bodichon*, p. 139.
11. Barbara Leigh Bodichon, *Algeria Considered as a Winter Residence for the English* (London: English Women's Journal, 1858), p. 104.
12. Bodichon, *Algeria Considered as a Winter Residence*, pp. 104–5.
13. George W. Harris, *'The' Practical Guide to Algiers* (London: George Philip, 1890), p. 3.
14. Harris, *'The' Practical Guide to Algiers*, p. 4.
15. Harris, *'The' Practical Guide to Algiers*, p. 7.
16. Harris, *'The' Practical Guide to Algiers*, p. 12.
17. The single major debate about choices in visions for the colony was between advocates of military and civilian rule (aside from the question of retaining or abandoning Algeria), and there was a certain symmetry between early French Algeria and the newly independent Algeria after 1962, for in both cases the admittedly critical question of what role the army would play in governing the new state also masked a set of questions about ideological direction which were to lie unanswered.
18. Hamdan Khodja, *Le Miroir: Aperçu historique et statistique sur la Régence d'Alger* (Paris: Sindbad, 1985), pp. 230–31.
19. Khodja, *Le Miroir*, p. 236.
20. Khodja, *Le Miroir*, p. 237.
21. Cooter et al., *War, Medicine and Modernity*, p. 11.
22. My analysis here connects with the growing field of studies which consider the 'complicity' of nineteenth-century feminism with the imperial project, such as Julia Clancy-Smith, 'The "Passionate Nomad" Reconsidered: A European Woman in L'Algérie Française (Isabelle Eberhardt, 1877–1904)', 61–78, in the important collection edited by Nupur Chaudhuri and Margaret Strobel, *Western Women and Imperialism: Complicity and Resistance* (Bloomington: Indiana University Press, 1992). A second important collection is Clare Midgley (ed.), *Gender and Imperialism* (Manchester: Manchester University Press, 1998).
23. Olivier Le Cour Grandmaison, *Coloniser. Exterminer: Sur la guerre et l'état colonial* (Paris: Fayard, 2005), pp. 8–9.
24. Gill, *The Language of French Orientalist Painting*, pp. 12–13.
25. William B. Cohen, *The French Encounter with Africans: White Responses to Blacks, 1530–1880* (Bloomington: Indiana University Press, 1980), p. 270.
26. Charles-Robert Ageron, *France Coloniale ou Parti Colonial?* (Paris: PUF, 1978), p. 19.
27. Léonard, *Les Medecins de l'ouest*, p. 1280.
28. Eugène Bodichon, *Aux Électeurs de l'Algérie* (Algiers: A. Bourget, 1848), p. 2.

29. Émile Temime, *Un Rêve méditerranéen: Des saint-simoniens aux intellectuels des années trente (1832–1962)* (Arles: Actes Sud, 2002).
30. Morsy, Magali (ed.), *Les Saint-Simoniens et l'orient: Vers la modernité* (Aix-en-Provence: Édisud, 1990), p. 14.
31. Magali, *Les Saint-Simoniens et l'orient*, p. 8.
32. Magali, *Les Saint-Simoniens et l'orient*, p. 9.
33. Magali, *Les Saint-Simoniens et l'orient*, p. 9.
34. Grove, *Green Imperialism*, especially p. 3.
35. Grove, *Green Imperialism*, p. xi.
36. Grove, *Green Imperialism*, p. 13.
37. Grove, *Green Imperialism*, p. 9.
38. Grove, *Green Imperialism*, p. 9.
39. Grove, *Green Imperialism*, p. 13.
40. Grove, *Green Imperialism*, p. 13.
41. Davis, Diana K., *Resurrecting the Granary of Rome: Environmental History and the French Colonial Expansion in North Africa* (Ohio: Ohio University Press, 2007).
42. Grove, *Green Imperialism*, p. 13.
43. Rozet, *Voyage dans la régence d'Alger*, II, p. 303.
44. Bodichon, *Hygiène à suivre en Algérie: Acclimatement des Européens*, pp. 5–6.
45. Bodichon, *Hygiène à suivre en Algérie: Acclimatement des Européens*, p. 9.
46. Dr Marcailhou d'Aymeric, *Manuel Hygiènique du Colon Algérien* (Algiers: Juillet Saint-Lager, 1874), pp. 29–30.
47. Michael A. Osborne, 'Resurrecting Hippocrates: Hygienic Sciences and the French Scientific Expeditions to Egypt, Morea and Algeria', in David Arnold (ed.), *Warm Climates and Western Medicine: The Emergence of Tropical Medicine 1500–1800* (Amsterdam: Rodopi, 1996), 80–98, p. 86.
48. Osborne, 'Resurrecting Hippocrates', p. 89.
49. Osborne, 'Resurrecting Hippocrates', p. 93.
50. Le Docteur Bodichon, *Le Vade Mecum de la politique française* (Alger: A. Bouyer, 1883), p. 3.
51. Bodichon, *Le Vade Mecum*, p. 8.
52. Bodichon, *Le Vade Mecum*, p. 9.
53. Bodichon, *Hygiène à suivre en Algérie: Hygiène Morale*, p. 5.
54. Bodichon, *Hygiène à suivre en Algérie: Hygiène Morale*, p. 5.
55. Bodichon, *Hygiène à suivre en Algérie: Hygiène Morale*, pp. 6, 9.
56. Bodichon, *Hygiène à suivre en Algérie: Hygiène Morale*, p. 11.
57. Bodichon, *Hygiène à suivre en Algérie: Hygiène Morale*, p. 11.
58. Bodichon, *Hygiène à suivre en Algérie: Hygiène Morale*, pp. 11, 12, 15.
59. Bodichon, *Hygiène à suivre en Algérie: Hygiène Morale*, p. 15.
60. Lorcin, *Imperial Identities*, p. 39.
61. Lorcin, *Imperial Identities*, pp. 123–24.
62. Lorcin, *Imperial Identities*, p. 144.
63. David Prochaska, *Making Algeria French: Colonialism in Bône 1870–1920* (Cambridge: Cambridge University Press, 1990).
64. Bodichon, *Hygiène à suivre en Algérie: Hygiène Morale*, p. 21.
65. Bodichon, *Hygiène à suivre en Algérie: Hygiène Morale*, p. 14.
66. *La France: doit-elle conserver Alger?* par un auditeur au conseil d'état (Paris: Bethuné et Plon, 1835), p. 66.
67. Ernest Mercier, *L'Algérie en 1880* (Paris: Challamel Aîné, 1880), p. 280.
68. Bodichon, *Hygiène à suivre en Algérie: Hygiène Morale*, p. 18.

69. Michael G. Vann, 'The Good, the Bad and the Ugly: Variation and Difference in French Racism in Colonial Indochine', in Sue Peabody and Tyler Stovall (eds), *The Color of Liberty: Histories of Race in France* (Durham: Duke University Press, 2003), pp. 187–205, p. 188.
70. Jacques Cohen, *Les Israélites de l'Algérie et le décret Crémieux* (Paris: Arthur Rousseau, 1900), pp. ix–x.
71. Béquet, *Répertoire du droit administratif*, II, p. 64.
72. Françoise Banat-Berger et al. (eds), *La Justice en Algérie 1830–1962* (Paris: La documentation Française, 2005), p. 341.
73. Béquet, *Répertoire du droit administratif*, II, p. 70.
74. Richard Fogarty and Michael Osborne, 'Constructions and Functions of Race in French Military Medicine, 1830–1920', in Sue Peabody and Tyler Stovall (eds), *The Color of Liberty: Histories of Race in France* (Durham: Duke University Press, 2003), pp. 206–36, pp. 225–26.
75. Lorcin, *Imperial Identities*, p. 9.
76. Prochaska, *Making Algeria French*, pp. 163–64, 232.
77. Prochaska, *Making Algeria French*, pp. 233–34.
78. Prochaska, *Making Algeria French*, p. 236.
79. Léonard, *Les Medecins de l'ouest*, p. 1280.
80. Lorcin, *Imperial Identities*, p. 3.
81. Lorcin, *Imperial Identities*, pp. 122–23.
82. Lorcin, *Imperial Identities*, pp. 122–23.
83. Bodichon, *Considérations sur l'Algérie*, p. 98.
84. Bodichon, *Le Vade Mecum*, p. 13.
85. Georges Renauld, *Adolphe Crémieux: Homme d'état français, juif et franc-maçon – le combat pour la république* (Paris: Detrad, 2002), p. 7.
86. Cohen, *Les Israélites de l'Algérie*, p. 222.
87. Cohen, *Les Israélites de l'Algérie*, pp. 224–32.
88. Raoul Bergot, *L'Algérie telle qu'elle est*, 2nd edn (Paris: Albert Savine, 1890), p. 207.
89. Bergot, *L'Algérie telle qu'elle est*, p. 205.
90. Albert Hugues, *La Nationalité française chez les musulmans de l'Algérie* (Paris: A. Chevalier-Marescq, 1899), pp. 188–89.

4 On Extermination

1. Un ancien capitaine de Zouaves, *Les Grottes de Dahara: Récit historique* (Paris: M. Blot, 1864), p. 6.
2. Le Cour Grandmaison, *Coloniser. Exterminer*, pp. 18–19.
3. Le Cour Grandmaison, *Coloniser. Exterminer*, pp. 18–19.
4. E. Plunkett, *The Past and Future of the British Navy* (London: Longman, 1846), p. 182.
5. Lord Percival Barton, *Algiers, with Notices of the Neighbouring States of Barbary*, 2 vols (London: Whittaker & Co., 1835), pp. 189–90.
6. Pierre Cœur, *L'Assimilation des indigènes musulmans* (Paris: A. Guédan, 1890), p. 33.
7. Barton, *Algiers*, p. 83.
8. Barton, *Algiers*, p. 83.
9. Barton, *Algiers*, p. 83.
10. Barton, *Algiers*, p. v.
11. René de Grieu, *Le Duc d'Aumale et l'Algérie* (Paris: Blériot et Gautier, 1884), p. 15.

12. Maréchal Clauzel, *Nouvelles observations de M. le Maréchal Clauzel sur la colonisation d'Alger addressées à M. le Maréchal, ministre de la guerre, président du conseil* (Paris: Imprimerie Selligue, 1833), p. 2.
13. Clauzel, *Nouvelles observations*, pp. 3–4.
14. H. Limbourg, *Le Duc d'Aumale et sa deuxième campagne d'Afrique* (février à septembre 1841) (Paris: Édition spéciale de la revue hebdomadaire, 1915), p. 4.
15. Limbourg, *Le Duc d'Aumale*, p. 4.
16. Alain Corbin, 'Douleurs, souffrances et misères du corp', in Alain Corbin, Jean-Jacques Courtine, Georges Vigarello (eds), *Histoire du Corps, 2 vols, vol. II* (Paris: Éditions du Seuil, 2005), pp. 215–73, p. 215.
17. Corbin, 'Douleurs, souffrances', p. 215.
18. Corbin, 'Douleurs, souffrances', pp. 221–23.
19. Le Cour Grandmaison, *Coloniser. Exterminer*, p. 20.
20. Le Cour Grandmaison, *Coloniser. Exterminer*, pp. 21–22.
21. Khodja, *Le Miroir*, p. 25.
22. Khodja, *Le Miroir*, p. 32.
23. Khodja, *Le Miroir*, pp. 214–15.
24. H. d'Ideville, *Les Petits côtés de l'histoire: notes intimes et documents inédits, 1870–1884* (Paris: Calmann Lévy, 1884), pp. 156–57.
25. Khodja, *Le Miroir*, pp. 38, 155, 211.
26. Khodja, *Le Miroir*, p. 27.
27. Khodja, *Le Miroir*, p. 213.
28. Khodja, *Le Miroir*, p. 215.
29. Khodja, *Le Miroir*, p. 264.
30. H. d'Ideville, *Memoirs of Marshal Bugeaud: From his Private Correspondence and Original Documents, 1784–1849, 2 vols* (London: Hurst and Blackett, 1884), p. 91.
31. d'Ideville (ed.), *Memoirs of Marshal Bugeaud*, p. 91.
32. d'Ideville (ed.), *Memoirs of Marshal Bugeaud*, pp. 91–92.
33. d'Ideville (ed.), *Memoirs of Marshal Bugeaud*, p. 92.
34. d'Ideville (ed.), *Memoirs of Marshal Bugeaud*, p. 93.
35. d'Ideville (ed.), *Memoirs of Marshal Bugeaud*, p. 92.
36. Edward Behr, *The Algerian Problem* (London: Hodder and Stoughton, 1961), p. 23.
37. A. Desjobert, *La Question d'Alger: Politique, Colonisation, Commerce* (Paris: A. Dufart, 1837); *L'Algérie en 1838* (Paris: P. Dufart, 1838); *L'Algérie en 1844* (Paris: Guillaumin, 1844); *L'Algérie en 1846* (Paris: Guillaumin, 1846).
38. Desjobert, *L'Algérie en 1838*, p. 52.
39. Desjobert, *L'Algérie en 1838*, p. 51.
40. Desjobert, *L'Algérie en 1838*, p. 51.
41. Desjobert, *La Question d'Alger*, pp. 92–93.
42. Le Cour Grandmaison, *Coloniser. Exterminer*, pp. 18–19, 118–19.
43. Desjobert, *L'Algérie en 1838*, pp. 85–86.
44. Desjobert, *La Question d'Alger*, p. 84.
45. Desjobert, *La Question d'Alger*, p. 85.
46. Desjobert, *La Question d'Alger*, p. 86.
47. Desjobert, *La Question d'Alger*, pp. 89–90.
48. Desjobert, *L'Algérie en 1838*, pp. 51–52.
49. Desjobert, *L'Algérie en 1844*, p. 42.
50. Desjobert, *L'Algérie en 1844*, p. 42.
51. Renault, *Première lettre à M. Passy*, p. 12.
52. Renault, *Première lettre à M. Passy*, p. 12.

53. M. Fumeron d'Ardeuil, *Observations sur la situation et l'avenir de nos possessions d'afrique* (Paris: Belin, 1839), p. 12.
54. Ardeuil, *Observations sur la situation*, p. 12.
55. *Le Moniteur Universel*, 68, samedi 9 mars 1833, p. 647.
56. *Le Moniteur Universel*, p. 647.
57. *Le Moniteur Universel*, p. 648.
58. *Appel en faveur d'Alger et de l'Afrique du nord par un anglais* (Paris: Dondet-Dupré, 1833), p. 3.
59. *Appel en faveur*, p. 9.
60. *Appel en faveur*, pp. 20–21.
61. Desjobert, *La Question d'Alger*, p. 90.
62. Desjobert, *La Question d'Alger*, p. 90.
63. J. de S.-D., 'Quelques considérations sur le projet d'expulser les Ottomans de l'Europe et de les remplacer par une population gréco-slave', *Revue de l'orient: Bulletin de la société orientale*, 6 (1845), 286–93.
64. Desjobert, *L'Algérie en 1838*, p. 52.
65. Desjobert, *La Question d'Alger*, pp. 94–95.
66. Desjobert, *La Question d'Alger*, p. 87.
67. Desjobert, *L'Algérie en 1838*, p. 57.
68. Desjobert, *L'Algérie en 1844*, p. 46.
69. Desjobert, *L'Algérie en 1846*, p. 46.
70. Desjobert, *L'Algérie en 1846*, p. 46.
71. Warnery, *Résumé de la situation*, p. 2.
72. Le Pays de Bourjolly, *Projets sur l'Algérie*, pp. 100–101.
73. Frantz Fanon, *L'An V de la revolution algérienne*, 2nd edn (Paris: François Maspéro, 1960), p. 119.
74. Le Cour Grandmaison, *Coloniser. Exterminer*, pp. 114–15.
75. Desjobert, *L'Algérie en 1844*, p. 44–45.
76. A. Mattei, *Protestation contre les détracteurs du système administratif suivi actuellement en Algérie/Coup d'œuil sur les différentes dominations en Afrique* (Paris: E. Dentu, 1869), p. 7.
77. Cœur, *L'Assimilation des indigènes musulmans*, p. 82.
78. Cœur, *L'Assimilation des indigènes musulmans*, p. 81.
79. Henri de Sarrauton, *La Question algérienne* (Oran: Paul Perrier, 1891).
80. de Sarrauton, *La Question algérienne*, p. 3.
81. de Sarrauton, *La Question algérienne*, p. 4.
82. de Sarrauton, *La Question algérienne*, pp. 5–6.
83. de Sarrauton, *La Question algérienne*, p. 6.
84. de Sarrauton, *La Question algérienne*, p. 27.
85. Desjobert, *L'Algérie en 1844*, p. 43.
86. Un ancien Capitaine de Zouaves, *Les Grottes du Dahara: Récit historique* (Paris: M. Blot. 1864), p. 5.
87. Un ancien Capitaine de Zouaves, *Les Grottes du Dahara*, pp. 5–6.
88. Un ancien Capitaine de Zouaves, *Les Grottes du Dahara*, p. 6.
89. Un ancien Capitaine de Zouaves, *Les Grottes du Dahara*, p. 9.
90. Un ancien Capitaine de Zouaves, *Les Grottes du Dahara*, p. 7.
91. Un ancien Capitaine de Zouaves, *Les Grottes du Dahara*, p. 9.
92. Un ancien Capitaine de Zouaves, *Les Grottes du Dahara*, p. 9.
93. Un ancien Capitaine de Zouaves, *Les Grottes du Dahara*, p. 9.
94. Un ancien Capitaine de Zouaves, *Les Grottes du Dahara*, p. 9.

95. Un ancien Capitaine de Zouaves, *Les Grottes du Dahara*, p. 10.
96. Un ancien Capitaine de Zouaves, *Les Grottes du Dahara*, p. 10.
97. Général Derrécagaix, *Le Maréchal Pélissier, Duc de Malakoff* (Paris: Librairie Militaire R. Chapelot, 1911), p. 181.
98. Derrécagaix, *Le Maréchal Pélissier*, p. 180.
99. Derrécagaix, *Le Maréchal Pélissier*, p. 180.
100. Derrécagaix, *Le Maréchal Pélissier*, p. 190.
101. Derrécagaix, *Le Maréchal Pélissier*, p. 181.
102. Henry C. Wright, *Defensive War Proved to be a Denial of Christianity and of the Government of God* (London: Charles Gilpin, 1846), p. 39.
103. Wright, *Defensive War Proved to be a Denial of Christianity*, p. 39.
104. Derrécagaix, *Le Maréchal Pélissier*, p. 192; Commandant Grandin, Léonce, *Le maréchal Pélissier, Duc de Malakoff* (Abbeville: C. Paillart, 1901)., p. 64.
105. Bodichon, *Considérations sur l'Algérie*, p. 35. See also Desjobert, *L'Algérie en 1838*, p. 56.
106. Derrécagaix, *Le Maréchal Pélissier*, p. 174.
107. Derrécagaix, *Le Maréchal Pélissier*, p. 182.
108. Derrécagaix, *Le Maréchal Pélissier*, p. 182.
109. Derrécagaix, *Le Maréchal Pélissier*, p. 184.
110. Derrécagaix, *Le Maréchal Pélissier*, p. 185.
111. Derrécagaix, *Le Maréchal Pélissier*, p. 185.
112. Julie de Marguerittes, *Italy and the War of 1859* (Philadelphia: George G. Evans, 1859), p. 130.
113. Derrécagaix, *Le Maréchal Pélissier*, p. 185.
114. Derrécagaix, *Le Maréchal Pélissier*, p. 185.
115. Le Cour Grandmaison, *Coloniser. Exterminer*, p. 8
116. Derrécagaix, *Le Maréchal Pélissier*, p. 186.
117. Derrécagaix, *Le Maréchal Pélissier*, p. 186.
118. Derrécagaix, *Le Maréchal Pélissier*, p. 186.
119. Derrécagaix, *Le Maréchal Pélissier*, p. 186.
120. Derrécagaix, *Le Maréchal Pélissier*, p. 187.
121. Derrécagaix, *Le Maréchal Pélissier*, p. 187.
122. Derrécagaix, *Le Maréchal Pélissier*, p. 187.
123. Derrécagaix, *Le Maréchal Pélissier*, p. 188.
124. Derrécagaix, *Le Maréchal Pélissier*, p. 189.
125. Un ancien Capitaine de Zouaves, *Les Grottes du Dahara*, p. 10.
126. Behr, *The Algerian Problem*, p. 23.
127. Derrécagaix, *Le Maréchal Pélissier*, pp. 183, 189.
128. Derrécagaix, *Le Maréchal Pélissier*, p. 190.
129. Derrécagaix, *Le Maréchal Pélissier*, p. 190.
130. Le Cour Grandmaison, *Coloniser. Exterminer*, p. 117.
131. P. Marbaud, *Coup d'œil sur l'Algérie pendant la crise de 1859–1860 et réflexions sur le décret relatif à la vente des terres domaniales* (Constantine: V. Guende, 1860), pp. 5–7.
132. Marbaud, *Le Maréchal Pelissier*, pp. 6–7.
133. Grandin, *Le Maréchal Pelissier*, p. 67.
134. Arsène Berteuil, *L'Algérie française: histoires, mœurs, coutûmes, industrie, agriculture* (Paris: E. Dentu, 1856), p. 297.
135. Un ancien Capitaine de Zouaves, *Les Grottes du Dahara*, p. 10.
136. Un ancien Capitaine de Zouaves, *Les Grottes du Dahara*, pp. 10–11.

137. Un ancien Capitaine de Zouaves, *Les Grottes du Dahara*, p. 11.
138. Un ancien Capitaine de Zouaves, *Les Grottes du Dahara*, p. 11.
139. Un ancien Capitaine de Zouaves, *Les Grottes du Dahara*, p. 12–13.
140. Un ancien Capitaine de Zouaves, *Les Grottes du Dahara*, p. 16.
141. Un ancien Capitaine de Zouaves, *Les Grottes du Dahara*, p. 16.
142. Derrécagaix, *Le Maréchal Pélissier*, pp. 194–95.
143. Derrécagaix, *Le Maréchal Pélissier*, p. 174.
144. G. Cluseret, *Mémoires du général Cluseret* (Paris: Lévy, 1887), p. 196.
145. de Marguerittes, *Italy and the War of 1859*, p. 128.
146. Wright, *Defensive War Proved to be a Denial of Christianity*, pp. 40–41.
147. Behr, *The Algerian Problem*, p. 23.
148. Richard Norman, *Ethics, Killing and War* (Cambridge: Cambridge University Press, 1995) , p. 188.
149. Thomas Nagel, *Mortal Questions* (Cambridge: Cambridge University Press, 1979), p. 54.
150. Behr, *The Algerian Problem*, pp. 24, 21.
151. Marcel Reggui, *Les massacres de Guelma: Algérie, mai 1945; une enquête inédite sur la furie des milices coloniales* (Paris: La Découverte, 2006).
152. James McDougall, 'Martyrdom and destiny: the inscription and imagination of Algerian history', in Ussama Makdisi and Paul A. Silverstein (eds), *Memory and Violence in the Middle East and North Africa* (Bloomington: Indiana University Press, 2006), 50–72, p. 50.
153. Silverstein, Paul A. and Makdisi, Ussama, 'Introduction: Memory and Violence in the Middle East and North Africa', in Ussama Makdisi and Paul A. Silverstein (eds), *Memory and Violence in the Middle East and North Africa* (Bloomington: Indiana University Press, 2006), pp. 1–24, p. 1.
154. John T. Parry, 'Pain, Interrogation, and the Body: State Violence and the Law of Torture', in John T. Parry (ed.), *Evil, Law and the State: Perspectives on State Power and Violence* (Amsterdam: Rodopi, 2006), 1–16, p. 1.
155. Benjamin Stora, 'The Algerian War in French memory: vengeful memory's violence', in Ussama Makdisi, and Paul A. Silverstein (eds), *Memory and Violence in the Middle East and North Africa* (Bloomington: Indiana University Press, 2006), pp. 151–74, p. 158.

5 On Attendance to Suffering and Demographic Collapse

1. CAOM, ALG 1 1/I/24.
2. Ageron, *Politiques Coloniales au Maghreb*, p. 228.
3. Annie Rey-Goldzeiguer, *Le Royaume arabe: la politique algérienne de Napoléon III, 1861–1870* (Algiers: SNED, 1977), p. 483.
4. David Arnold, *Famine: Social Crisis and Historical Change* (Oxford: Basil Blackwell, 1988), p. 124.
5. Arnold, *Famine*, p. 124.
6. Rey-Goldzeiguer, *Le Royaume arabe*, p. 444.
7. Rey-Goldzeiguer, *Le Royaume arabe*, p. 451.
8. CAOM, ALG 1I/15–16.
9. CAOM, ALG 1I/15–16.
10. CAOM, ALG 1I/15–16.
11. CAOM, ALG 1I/15–16.
12. CAOM, ALG 1I/15–16.

13. CAOM, ALG 1I/15–16.
14. CAOM, ALG 1I/15–16.
15. CAOM, ALG 1I/15–16.
16. Rey-Goldzeiguer, *Le Royaume arabe*, p. 448.
17. Rey-Goldzeiguer, *Le Royaume arabe*, p. 448.
18. CAOM, ALG 1I/15–16.
19. Magali Morsy, *North Africa 1800–1900: A Survey from the Nile Valley to the Atlantic* (London: Longman, 1984), p. 162.
20. Arnold, *Famine*, p. 120.
21. Arnold, *Famine*, p. 122.
22. Arnold, *Famine*, p. 126.
23. Arnold, *Famine*, p. 126.
24. Arnold, *Famine*, p. 126.
25. Desjobert, *L'Algérie en 1846*, p. 17.
26. Desjobert, *L'Algérie en 1846*, p. 17.
27. Desjobert, *L'Algérie en 1838*, p. 93.
28. Desjobert, *L'Algérie en 1846*, p. 16.
29. Desjobert, *L'Algérie en 1846*, p. 17.
30. Ageron, *Politiques Coloniales au Maghreb*, p. 74.
31. Ageron, *Politiques Coloniales au Maghreb*, p. 74.
32. Ageron, *Politiques Coloniales au Maghreb*, p. 74.
33. Lady Herbert, *A Search after Sunshine: Algeria in 1871* (London: Richard Bentley, 1872), pp. 96–97.
34. Ageron, *Politiques Coloniales au Maghreb*, p. 75.
35. Ageron, *Politiques Coloniales au Maghreb*, p. 76.
36. Ageron, *Politiques Coloniales au Maghreb*, pp. 74–75.
37. Ageron, *Politiques Coloniales au Maghreb*, p. 75.
38. Joseph Variot, *Les Pères Blancs ou Missionaires d'Alger* (Lille: Société St Augustin, 1887), p. 3.
39. Morsy, *North Africa 1800–1900*, p. 160.
40. François Renault, *Cardinal Lavigerie: Churchman, Prophet and Missionary* (London: Athlone Press, 1992), p. 94.
41. Renault, *Cardinal Lavigerie*, p. 95.
42. Renault, *Cardinal Lavigerie*, p. 95.
43. Gérard Bedel, *Sous la bannière du Sacré-Cœur: Le général de Sonis – Miles Christi* (Tournon-Saint-Martin, 1997), p. 84.
44. Aylward Shorter, *Cross and Flag in Africa: The 'White Fathers' during the Colonial Scramble* (Maryknoll: Orbis, 2006), p. 41.
45. Shorter, *Cross and Flag in Africa*, p. 145.
46. Paul Lesourd, *Les Pères Blancs du Cardinal Lavigerie* (Orléans: Bernard Grasset, 1935), p. 64.
47. Shorter, *Cross and Flag in Africa*, p. 145.
48. Variot, *Les Pères Blancs*, p. 7.
49. Renault, *Cardinal Lavigerie*, p. 88.
50. Renault, *Cardinal Lavigerie*, pp. 88–89.
51. Mike Davis, *Late Victorian Holocausts: El Niño Famines and the Making of the Third World* (London: Verso, 2001), p. 6.
52. Davis, *Late Victorian Holocausts*, p. 6.
53. Davis, *Late Victorian Holocausts*, p. 105.
54. Davis, *Late Victorian Holocausts*, p. 106.

55. Rey-Goldzeiguer, *Le Royaume arabe*, p. 474.
56. Rey-Goldzeiguer, *Le Royaume arabe*, p. 489.
57. Rey-Goldzeiguer, *Le Royaume arabe*, p. 489.
58. Auguste Dupré, *Lettres sur l'Algérie* (Bordeaux: G. Gounouilhou, 1870), pp. 25–26.
59. Eugène Jung, *L'Islam et les Musulmans dans l'Afrique du Nord* (Paris: Jeune Parque, 1930), p. 36.
60. Jung, *L'Islam et les Musulmans*, p. 36.
61. Paul Bourde, *A Travers l'Algérie: Souvenirs de l'excursion parlementaire* (septembre–octobre 1879) (Paris: G. Charpentier, 1880), pp. 162–63.
62. Bourde, *A Travers l'Algérie*, pp. 162–63.
63. Rey-Goldzeiguer, *Le Royaume arabe*, p. 491.
64. Paul Delavigne, Auguste Warnier , O. MacCarthy, U. Ranc, and J.A. Serpolet *Chemin de fer de l'Algérie par la ligne centrale du Tell* (Algiers: Beau, 1854), p. 63.
65. Delavigne and Warnier, *Chemin de fer*, pp. 90, 91.
66. Delavigne and Warnier, *Chemin de fer*, p. 91.
67. Delavigne and Warnier, *Chemin de fer*, p. 91.
68. Cherry, *Beyond the Frame*, p. 76.
69. Saint-Hypolite, *De l'Algérie: Moyens d'affermir nos possessions en 1840* (Paris: Bourgogne et Martinet, 1840), p. 5.
70. Douglas Porch, *The Conquest of the Sahara* (Oxford: Oxford University Press, 1986), p. 42.
71. Davis, *Late Victorian Holocausts*, p. 10.
72. Paul Bonnet, Étude sur le choléra: épidémies 1865, 1866, 1867 observées à Alger, doctoral thesis in medicine, Faculté de Médecine de Paris, Paris, 1870, pp. 41–42.
73. Grove, *Green Imperialism*, p. 7.
74. Bodichon, *Hygiène à suivre en Algérie: Hygiène Morale*, p. 23.
75. Ageron, *Politiques Coloniales au Maghreb*, p. 228.
76. Ageron, *Politiques Coloniales au Maghreb*, p. 228.
77. Ageron, *Politiques Coloniales au Maghreb*, p. 229.
78. d'Ideville, *Memoirs of Marshal Bugeaud*, p. 92.
79. d'Ideville, *Memoirs of Marshal Bugeaud*, p. 92.
80. Clancy-Smith, 'The Shaykh and his Daughter', pp. 119–36.
81. Jung, *L'Islam et les Musulmans*, p. 37.
82. Dupré, *Lettres sur l'Algérie*, p. 72.
83. Jung, *L'Islam et les Musulmans*, p. 29.
84. A. Warnier, *L'Algérie devant le sénat* (Paris: Challamel Ainé, 1863), p. 33.
85. Warnier, *L'Algérie devant le sénat*, p. 63.
86. Warnier, *L'Algérie devant le sénat*, p. 63.
87. Warnier, *L'Algérie devant le sénat*, p. 34.
88. *Le Moniteur Universel*, p. 648.
89. *Le Moniteur Universel*, p. 649.
90. *Le Moniteur Universel*, p. 649.
91. Marc Ferro, 'La conquête de l'Algérie', in Marc Ferro (ed.), *Le livre noir du colonialisme: XVIe – XXIe siècle: de l'extermination à la repentance* (Paris: Robert Laffont, 2003), pp. 490–505, p. 496.
92. Joost Van Vollenhoven, *Essai sur le fellah algérien* (Paris: Arthur Rousseau, 1903).
93. Van Vollenhoven, *Essai sur le fellah algérien*, p. 189.
94. Van Vollenhoven, *Essai sur le fellah algérien*, p. 190.
95. Van Vollenhoven, *Essai sur le fellah algérien*, p. 213.
96. Van Vollenhoven, *Essai sur le fellah algérien*, p. 213.

97. Van Vollenhoven, *Essai sur le fellah algérien*, pp. 213–14.
98. Van Vollenhoven, *Essai sur le fellah algérien*, p. 198.
99. Van Vollenhoven, *Essai sur le fellah algérien*, pp. 195–96.
100. Van Vollenhoven, *Essai sur le fellah algérien*, pp. 195–96.
101. Van Vollenhoven, *Essai sur le fellah algérien*, pp. 221–22.
102. Un chef de Bureau Arabe, *L'Algérie Assimilée: Étude sur la constitution et la réorganisation de l'Algérie* (Constantine: L. Marle, 1871).
103. Un chef de Bureau Arabe, *L'Algérie Assimilée*, pp. 7–8.
104. Un chef de Bureau Arabe, *L'Algérie Assimilée*, p. 9.
105. Louis Rinn, *Histoire de l'Insurrection de 1871 en Algérie* (Algiers: Librairie Adolphe Jourdan, 1891).
106. Rinn, *Histoire de l'Insurrection de 1871*, p. 1.
107. F. Leblanc de Prébois, *Bilan du régime civil de l'Algérie à la fin de 1871* (Paris: E. Dentu, 1872), pp. 15–16.
108. Leblanc de Prébois, *Bilan du régime civil*, p. 3.
109. Un ancien officier de l'armée d'Afrique, *L'Algérie devant l'Assemblée Nationale: causes des insurrections algériennes* (Versailles: Muzard, 1871), p. 1.
110. Un ancien officier de l'armée d'Afrique, *L'Algérie devant l'Assemblée Nationale*, p. 1.
111. Louis Serre, *Les Arabes Martyres: Étude sur l'insurrection de 1871 en Algérie* (Paris: E. Lachaud, 1873), p. 27.
112. Serre, *Les Arabes Martyres*, p. 28.
113. Un ancien officier de l'armée d'Afrique, *L'Algérie devant l'Assemblée Nationale*, p. 20.
114. Un ancien officier de l'armée d'Afrique, *L'Algérie devant l'Assemblée Nationale*, p. 21.
115. Un ancien officier de l'armée d'Afrique, *L'Algérie devant l'Assemblée Nationale*, p. 22.
116. E. Beauvois, En Colonne dans la Grand Kabyle: Souvenirs de l'insurrection de 1871 avec une relation du siége de Fort-National (Paris: Challamel, 1872), pp. 9, 11.
117. Serre, *Les Arabes Martyres*, p. 30.
118. Serre, *Les Arabes Martyres*, pp. 30–32.
119. Serre, *Les Arabes Martyres*, p. 32.
120. A. Ducrot, *La Vérité sur l'Algérie* (Paris: E. Dentu, 1871), p. 67.
121. Bourde, *A Travers l'Algérie*, p. 389.
122. Bourde, *A Travers l'Algérie*, p. 233.
123. Kamel Kateb, *La Statistique colonial en Algérie (1830–1962): Entre la reproduction du système métropolitain et les impératifs d'adaptation à la réalité algérienne*: http://www.insee.fr/fr/ffc/docs_ffc/cs112b.pdf
124. Rey-Goldzeiguer, *Le Royaume arabe*, pp. 457–58.
125. Dupré, *Lettres sur l'Algérie*, pp. 71–72.
126. Bourde, *A Travers l'Algérie*, p. 233.
127. Ferro, *Le livre noir du colonialisme*, p. 499.
128. Dupré, *Lettres sur l'Algérie*, p. 72.
129. Dupré, *Lettres sur l'Algérie*, Bordeaux, p. 71.
130. J. Ch. M. Boudin, *Histoire statistique de la colonisation et de la population en Algérie* (Paris: J.B. Baillière, 1853), p. 21.
131. Boudin, *Histoire statistique de la colonisation*, p. 21.
132. Boudin, *Histoire statistique de la colonisation*, p. 21.
133. Prochaska, *Making Algeria French*, pp. 142–44.
134. Prochaska, *Making Algeria French*, p. 144.
135. M. Bennoune, *The Making of Contemporary Algeria* (Cambridge: Cambridge University Press, 1988), p. 54.

136. Dr A. Warnier, *L'Algérie devant l'opinion publique: indigènes et immigrants, examination rétrospectif* (Paris: Challamel Ainé, 1864).
137. Warnier, *L'Algérie devant l'opinion publique*, pp. 28–29.
138. Warnier, *L'Algérie devant l'opinion publique*, p. 29.
139. Rey-Goldzeiguer, *Le Royaume arabe*, p. 11.
140. Ferro, *Le livre noir du colonialisme*, p. 497.
141. Dr René Ricoux, *La Démographie figurée de l'Algérie: étude statistique des populations européennes qui habitent l'Algérie* (Paris: G. Masson, 1880).
142. Ricoux, *La Démographie figurée de l'Algérie*, p. 260.
143. Ricoux, *La Démographie figurée de l'Algérie*, p. 261.
144. Ricoux, *La Démographie figurée de l'Algérie*, p. 261.
145. Ricoux, *La Démographie figurée de l'Algérie*, pp. 261–62.
146. Ricoux, *La Démographie figurée de l'Algérie*, p. 262.
147. Ricoux, *La Démographie figurée de l'Algérie*, p. 262.
148. Ricoux, *La Démographie figurée de l'Algérie*, p. 263.
149. Le Cour Grandmaison, *Coloniser. Exterminer*, p. 80.
150. Dupré, *Lettres sur l'Algérie*, pp. 10–11.
151. Rey-Goldzeiguer, *Le Royaume arabe*, p. 441.
152. Rey-Goldzeiguer, *Le Royaume arabe*, p. 457.
153. CAOM, ALG 10/I/8.
154. Rey-Goldzeiguer, *Le Royaume arabe*, p. 553.
155. Charles-André Julien, *Une pensée anticoloniale: positions 1914–1979* (Paris: Sindbad, 1979), pp. 42–52; Nouschi, *Enquête sur le niveau de vie des populations rurales constantinois*, pp. viii–ix.
156. Nouschi, *Enquête sur le niveau de vie des populations rurales constantinois*, p. ix.
157. Nouschi, *Enquête sur le niveau de vie des populations rurales constantinois*, p. ix.

6 On the Just and Sovereign Testimony of Abdel Kader ben Zahra

1. Letter of 30 May 1884, found in Centre des Archives d'Outre-Mer [hereafter CAOM], box ALG 113 bis, file 1u/1
2. Clancy-Smith, 'The Shaykh and his Daughter', pp. 145–63.
3. CAOM, ALG 113 bis, file 1u/1.
4. CAOM, ALG 113 bis, file 1u/1.
5. CAOM, ALG 113 bis, file 1u/1.
6. CAOM, ALG 113 bis, file 1u/1.
7. CAOM, ALG 113 bis, file 1u/1.
8. Prochaska, *Making Algeria French*, p. 233.
9. CAOM, ALG 113 bis, file 1u/1.
10. CAOM, ALG 113 bis, file 1u/1.
11. CAOM, ALG 113 bis, file 1u/1.
12. T. Pein, *Lettres familières sur l'Algérie, un petit royaume arabe* (Paris: Tanera, 1871), pp. 19–20.
13. CAOM, ALG 113 bis, file 1u/1.
14. CAOM, ALG 113 bis, file 1u/1.
15. CAOM, ALG 113 bis, file 1u/1.
16. CAOM, ALG 113 bis, file 1u/1.
17. CAOM, ALG 113 bis, file 1u/1.
18. CAOM, ALG 113 bis, file 1u/1.

7   On Injustice and the Disavowal of Autonomy

1. CAOM, ALG 113 bis, file 1u/139.
2. CAOM, ALG 113 bis, file 1u/139.
3. CAOM, ALG 113 bis, file 1u/139.
4. CAOM, ALG 113 bis, file 1u/68.
5. CAOM, ALG 113 bis, file 1u/68.
6. CAOM, ALG 113 bis, file 1u/68.
7. CAOM, ALG 113 bis, file 1u/68.
8. CAOM, ALG 113 bis, file 1u/28.
9. CAOM, ALG 113 bis, file 1u/28.
10. CAOM, ALG 113 bis, file 1u/28.
11. CAOM, ALG 113 bis, file 1u/28.
12. CAOM, ALG 113 bis, file 1u/28.
13. CAOM, ALG 113 bis, file 1u/28.
14. CAOM, ALG 113 bis, file 1u/28.
15. CAOM, ALG 113 bis, file 1u/28.
16. CAOM, ALG 113 bis, file 1u/28.
17. CAOM, ALG 113 bis, file 1u/28.
18. CAOM, ALG 113 bis, file 1u/28.
19. CAOM, ALG 113 bis, file 1u/28.
20. CAOM, ALG 113 bis, file 1u/138.
21. CAOM, ALG 113 bis, file 1u/138.
22. CAOM, ALG 113 bis, file 1u/138.
23. CAOM, ALG 113 bis, file 1u/138.
24. CAOM, ALG 113 bis, file 1u/138.
25. CAOM, ALG 113 bis, file 1u/138.
26. CAOM, ALG 113 bis, file 1u/140.
27. CAOM, ALG 113 bis, file 1u/140.
28. CAOM, ALG 113 bis, file 1u/140.
29. CAOM, ALG 113 bis, file 1u/140.
30. CAOM, ALG 113 bis, file 1u/140.
31. CAOM, ALG 113 bis, file 1u/140.
32. CAOM, ALG 113 bis, file 1u/140.
33. CAOM, ALG 113 bis, file 1u/140.
34. CAOM, ALG 113 bis, file 1u/140.
35. CAOM, ALG 113 bis, file 1u/140.
36. CAOM, ALG 113 bis, file 1u/140.
37. CAOM, ALG 113 bis, file 1u/140.
38. CAOM, ALG 113 bis, file 1u/140.
39. CAOM, ALG 113 bis, file 1u/140.
40. CAOM, ALG 113 bis, file 1u/140.
41. CAOM, ALG 113 bis, file 1u/140.
42. CAOM, ALG 113 bis, file 1u/140.
43. CAOM, ALG 113 bis, file 1u/140.
44. CAOM, ALG 113 bis, file 1u/140.
45. CAOM, ALG 113 bis, file 1u/140.
46. CAOM, ALG 113 bis, file 1u/140.
47. Sadek Sellam, *L'Islam et les musulmans en France: perceptions, craintes et réalités* (Paris: Tougui, 1987), p. 280.

48. Sellam, *L'Islam et les musulmans en France*, p. 281.
49. http://www.le-carrefour-de-lislam.com/Grenier/grenier1.htm – 16.07.06
50. Sellam, *L'Islam et les musulmans en France*, p. 292.
51. Sellam, *L'Islam et les musulmans en France*, p. 292.
52. M. B. Malfroy, Olivier, P. Bichet and J. Guiraud, *Histoire de Pontarlier* (Besançon: Cêtre, 1979), pp. 216–17.
53. Malfroy, Bichet and Guiraud, *Histoire de Pontarlier*, p. 216.
54. Sellam, *L'Islam et les musulmans en France*, p. 288.
55. Sellam, *L'Islam et les musulmans en France*, p. 289.
56. Sellam, *L'Islam et les musulmans en France*, p. 295.
57. Sellam, *L'Islam et les musulmans en France*, p. 283.
58. Sellam, *L'Islam et les musulmans en France*, p. 293.
59. Sellam, *L'Islam et les musulmans en France*, p. 293.
60. Sellam, *L'Islam et les musulmans en France*, p. 294.
61. Sellam, *L'Islam et les musulmans en France*, p. 294.
62. Sellam, *L'Islam et les musulmans en France*, p. 294.
63. Sellam, *L'Islam et les musulmans en France*, p. 294.
64. Sellam, *L'Islam et les musulmans en France*, p. 283.
65. Sellam, *L'Islam et les musulmans en France*, p. 294–95.

Bibliography

Abdel-Jouad, Hédi, *Rimbaud et l'Algérie* (Paris: Paris-Méditerranée, 2004).

Achi, Raberh, 'La laïcité à l'épreuve de la situation coloniale', in Banat-Berger Françoise (ed.), *La Justice en Algérie 1830–1962* (Paris: La documentation Française, 2005), 163–176.

Acot, Pascal (ed.), *The European Origins of Scientific Ecology*, 2 vols (Paris: EAC, 1998).

Adam, Hussein Mohaned, *The Social and Political Thought of Frantz Fanon*, Harvard University PhD thesis, 1974.

Adamson, Kay, *Political and Economic Thought and Practice in Nineteenth-Century France and the Colonization of Algeria* (Lewiston: Edwin Mellon, 2002).

Affaire de l'Oued-Mahouine (Cercle de Tébessa): Massacre d'une caravane (27 victimes), 2 vols (Constantine: L. Marle, 1870).

Ageron, Charles-Robert, *Les Algériens Musulmans et la France (1871–1919)* (Paris: PUF, 1968).

Ageron, Charles-Robert, *Politiques Coloniales au Maghreb* (Paris: PUF, 1972).

Ageron, Charles-Robert, *L'anticolonialisme en France de 1871 à 1914* (Paris: PUF, 1973).

Ageron, Charles-Robert, *France Coloniale ou Parti Colonial?* (Paris: PUF, 1978).

Ageron, Charles-Robert, *Histoire de l'Algérie contemporaine*, 8th edn (Paris: PUF, 1983).

Ageron, Charles-Robert, 'Peut-on parler d'une politique des "Royaumes Arabes" de Napoléon III?' in Morsy, Magali (ed.) *Les Saint-Simoniens et l'orient: Vers la modernité* (Aix-en-Provence: Édisud, 1990), 83–96.

Ageron, Charles-Robert, *De l'Algérie "française" à l'Algérie algérienne* (Paris: Editions Bouchène, 2005).

Ageron, Charles-Robert, *Genèse de l'Algérie algérienne* (Paris: Editions Bouchène, 2005).

Ageron, Charles-Robert, *Le gouvernement du général Berthezène à Alger en 1831* (Paris: Editions Bouchène, 2005).

Agnély, H. (c.1866) *Le climat de l'Algérie* (Algiers: J.B. Dubois).

Aisenberg, Andrew Robert, *Contagion: Disease, Government and the 'Social Question' in Nineteenth-Century France* (Stanford: Stanford University Press, 1999).

Aitken, William, *The Science and Practice of Medicine* (London: Lindsay and Blakiston, 1868).

Akhmisse, Mustapha, 'Histoire de la médecine au Maroc: des origines à l'avènement du Protectorat', *Histoire des Sciences Médicales*, 26 (1992), 263–269.

Alchon, Suzanne Austin, *Native Society and Disease in Colonial Ecuador* (Cambridge: Cambridge University Press, 1991).

Alexander, Martin S., Martin Evans and J.F.V. Keiger (eds), *The Algerian War and the French Army, 1954–62* (London: Palgrave Macmillan, 2002).

L'Algérie Assimilée: Étude sur la Constitution et la Réorganisation de l'Algérie, par un chef de Bureau Arabe (Constantine: L. Marle, 1871).

d'Allemagne, Henry-René, *Prosper-Enfantin et les Grandes Entreprises du XIXe siècle* (Paris: Librairie Grund, 1935).

Alvarez de Morales y Ruiz-Matas, Camilo (ed.), *'El Libro de la Almohada' de Ibn Wafid de Toledo (Recetario médico árabe del siglo xvi)* (Toledo: I.P.I.E.T., 1980).

Amrouche, Jean El-Mouhoub, *Un Algérien s'addresse aux Français ou l'histoire d'Algérie par les textes (1943–1961)* (Paris: Awal/Harmattan, 1994).

Amundsen, Darrell W., *Medicine, Society, and Faith in the Ancient and Medieval Worlds* (Baltimore: The Johns Hopkins University Press, 1996).

Appel en faveur d'Alger et de l'Afrique du nord par un anglais (Paris: Dondet-Dupré, 1833).

Archives de l'Institut Pasteur d'Algérie, 8–1, 1930.

Armand-Hain, Victor, *A la nation sur Alger* (Paris: Lachevardiere, 1832).

Arnold, David, *Famine: Social Crisis and Historical Change* (Oxford: Basil Blackwell, 1988).

Arnold, David, *Colonizing the Body: State Medicine and Epidemic Disease in Nineteenth-Century India* (Berkeley: University of California Press, 1993).

Arnold, David (ed.), *Warm Climates and Western Medicine: The Emergence of Tropical Medicine 1500–1800* (Amsterdam: Rodopi, 1996).

Arnold, David, *Imperial Medicine and Indigenous Societies* (Manchester: Manchester University Press, 1998).

Aussaresses, Paul, *Pour la France: services spéciaux 1942–1954* (Monaco: Rocher, 2001).

d'Aymeric, Dr Marcailhou, *Manuel Hygiènique du Colon Algérien* (Algiers: Juillet Saint-Lager, 1874).

Bahloul, Joelle, *The Architecture of Memory: A Jewish-Muslim Household in Colonial Algeria, 1937–1962* (Cambridge: Cambridge University Press, 1996).

Baker, Robert, 'The History of Medical Ethics', in Bynum, W.F. and Roy Porter (eds), *Encyclopedia of the History of Medicine*, 2 vols (London: Routledge, 1993), I, 852–887.

Banat-Berger, Françoise (ed.), *La Justice en Algérie 1830–1962* (Paris: La documentation Française, 2005).

de Bartillat, Le Marquis, *Aperçu sur la colonisation d'Alger* (Paris: Le Normant, 1837).

Barton, Lord Percival, *Algiers, with Notices of the Neighbouring States of Barbary*, 2 vols (London: Whittaker & Co., 1835).

Battandier, J.A. and Louis Trabut, *Algérie: Plantes Médicales: Essences et Parfums* (Algiers: Giralt, 1889).

Baude, Baron, *L'Algérie*, 2 vols (Paris: Arthus Bertrand, 1841).

Beauchamp, Tom L. and James F. Childress, *Principles of Biomedical Ethics*, 4th edn (Oxford: Oxford University Press, 1994).

Beauvois, E., *En Colonne dans la Grand Kabyle: Souvenirs de l'insurrection de 1871 avec une relation du siége de Fort-National* (Paris: Challamel, 1872).

Becker, Lawrence (ed.), *Encyclopedia of Ethics* (New York: Garland, 1992).

Bedel, Gérard, *Sous la bannière du Sacré-Cœur: Le général de Sonis – Miles Christi* (Tournon-Saint-Martin, 1997).

Bégué, Jean-Michel, *A Century of French Psychiatry in Algeria (1830–1939)* (Rotterdam: Erasmus Publishing, 1993).

Bégué, Jean-Michel, 'French Psychiatry in Algeria (1830–1962): From Colonial to Transcultural', in *History of Psychiatry*, special issue (1996), 533–548.

Behr, Edward, *The Algerian Problem* (London: Hodder and Stoughton, 1961).

Benhabilès, Chérif, *L'Algérie Française vue par un indigène* (Algiers: Orientale Fontana Frères, 1894).

Benjamin, Roger (ed.), *Renoir and Algeria* (Yale: Yale University Press, 2003).

Benjamin, Walter, *Charles Baudelaire: A Lyric Poet in the Era of High Capitalism*, trans. Zohn, Harry (London: Verso, 1983).

Benjamin, Walter, *Paris: capitale du XIXe siècle: Le Livre des Passages*, trans. Jean Lacoste (Paris: Les Éditions du Cerf, 1989).

Bennani, Jalil, *La psychanalyse au pays des saints: les débuts de la psychiatrie et de la psychanalyse au Maroc* (Casablanca: Éditions de Fennec, 1996).

Bennett, J. Henry, *Winter and Spring on the Shores of the Mediterranean* (London: John Churchill, 1870).

Bennoune, M. *The Making of Contemporary Algeria* (Cambridge: Cambridge University Press, 1988).

Bensimon-Donath, Doris, *Socio-demographie des juifs de France et d'Algérie: 1867–1907* (Paris: Publications Orientalistes de France, 1976).

Béquet, Léon and Marcel Simon, *Répertoire du droit administratif: Algérie – gouvernement, administration, législation*, 3 vols (Paris: Paul Dupont, 1883).

Berbrugger, Louis Adrien, *Algérie: Historique, Pittoresque et Monumentale* 5 vols (Paris: Delahaye, 1843).

Bergot, Raoul, *L'Algérie telle qu'elle est*, 2nd edn (Paris: Albert Savine, 1890).

Berteuil, Arsène, *L'Algérie française: histoires, mœurs, coutûmes, industrie, agriculture* (Paris: E. Dentu, 1856).

Berthelier, Robert, *L'homme maghrébin dans la littérature psychiatrique* (Paris: L'Harmattan, 1996).

Bertherand, A., 'De la prostitution en Algérie', in Parent-Duchatelet, A.J.B., *La Prostitution de Paris*, tome 2 (Paris: J.B. Baillière, 1857), 536–558.

Bertherand, A., *Campagnes de Kabylie: histoire médico-chirurgicale des expéditions de 1854, 1856 et 1857* (Paris: J.B. Baillière, 1862).

Bertherand, E.L., *Médecine et hygiène des Arabes* (Paris: Germer Baillière, 1855).

Bertherand, E.L., *Les chemins de fer au point de vue sanitaire* (Arbois: Javel, 1862).

Bertherand, E.L., *Hygiène musulmane*, 2nd edn (Paris: Challamel Ainé, 1874).

Bertherand, E.L., *Du suicide chez les musulmans de l'Algérie* (Paris: Challamel Ainé, 1875).

Bertherand, E.L., *Hygiène du colon en Algérie* (Paris: Challamel Ainé, 1875).

Bertrand-Cadi, Jean-Yves, *Le colonel Chérif Cadi: serviteur de l'Islam et de la République* (Paris: Maisonneuve & Larose, 2005).

Bewell, Alan, *Romanticism and Colonial Disease* (Baltimore: The Johns Hopkins University Press, 1999).

Bezombes, Louis, *Étude sur l'organisation de la justice française en Algérie depuis la conquête jusqu'à nos jours* (Philippeville: L. Denis Aîné, 1870).

Bidwell, Robin, *Morocco under Colonial Rule: French Administration of Tribal Areas, 1912 1956* (London: Frank Cass, 1973).

Blanc, Paul, *La vie de colon en Algérie* (Algiers: Juillet Saint-Lager, 1874).

Blanqui, M., *Algérie: Rapport sur la situation économique de nos possessions dans le nord de l'Afrique* (Paris: W.Coquebert, 1840).

Bodichon, Eugène, *Considérations sur l'Algérie* (Paris: Schneider & Legrand, 1845).

Bodichon, Eugène, *Études sur l'Algérie et l'Afrique* (Algiers: Bodichon, 1847).

Bodichon, Eugène, *Aux Électeurs de l'Algérie* (Algiers: A. Bourget, 1848).

Bodichon, Eugène, 'Projet d'une exploration politique, commerciale et scientifique d'Alger à Tombouctou par le Sahara', *Bulletin de la Société de Géographie*, juillet et août 1849.

Bodichon, Eugène, *Hygiène à suivre en Algérie: Acclimatement des Européens* (Algiers: Rey, Delavigne et Compagnie, 1851).

Bodichon, Eugène, *Hygiène à suivre en Algérie: Hygiène Morale* (Algiers: Rey, Delavigne et Compagnie, 1851).

Bodichon, Eugène, *De l'Humanité* (Algiers: A. Lacroix, 1852).

Bodichon Barbara Leigh, *Algeria Considered as a Winter Residence for the English* (London: English Women's Journal, 1858).

Bodichon, Le Docteur, *Le Vade Mecum de la politique française* (Alger: A. Bouyer, 1883).

Bohrer, M., aka Jean de Fermatou, *L'Algérie aux français d'origine ou le véritable sens de la question du decret Crémieux et sur des lois algériennes sur la naturalisation* (Algiers: Baldachino-Laronde-Viguier, 1898).

Bonah, Christian, ' "Experimental Rage": The Development of Medical Ethics and the Genesis of Scientific Facts – Ludwig Fleck: An Answer to the Crisis of Modern Medicine in Interwar Germany?' *Social History of Medicine*, 15–2 (2002), 187–207.

Bonnafont, Le Docteur, *Douze ans en Algérie* (Paris: Librairie de la Société des Gens et des Lettres, 1880).

Bonnet, Paul, *Étude sur le choléra: épidémies 1865, 1866, 1867 observées à Alger*, University of Paris, PhD thesis in medicine, faculty of medicine of Paris, 1870.

Bonzom, E., *Le Typhus Contagieux (Peste Bovine – Pestis Bovina)* (Algiers: Aillaud, 1871).

Borrer, Dawson, *Narrative of a Campaign against the Kabailes of Algeria with the Mission of M. Suchet to the Emir Abdel-Kader for an Exchange of Prisoners* (London: Longman, 1848).

Boucebci, M., *Psychiatrie: Société et Développement*, 2nd edn (Algiers: SNED, 1978).

Boudin, J. Ch. M., *Hygiène militaire comparée et statistique médicale des armées de terre et de mer* (Paris: J.B. Baillière, 1848).

Boulos, Loufty, *Medicinal Plants of North Africa* (Algonac: Reference Publications, 1983).

Bourde, Paul, *A Travers l'Algérie: Souvenirs de l'excursion* parlementaire (septembre–octobre 1879) (Paris: G. Charpentier, 1880).

Bourdieu, Pierre, *Ce que parler veut dire* (Paris: Fayard, 1982).

Bourguet, Marie-Noëlle, Bernard Lepetit, Daniel Nordman and Maroula Sinarellis (eds), *L'Invention scientifique de la Mediterranée: Egypte, Morée, Algérie* (Paris: Editions EHESS, 1998).

Bourouiba, Rachid (ed.), *La Construction du Maghreb – Actes du Colloque* (Tunis: Université de Tunis, 1983).

Bourquia, Rahma (ed.), 'Héritages culturels du Maghreb: histoire et mémoire', special issue of *Prologues: revue maghrébine du livre*, 29/30, (2004).

Boutaleb, Abdelkader, *L'Emir Abd-el-Kader et la formation de la nation algérienne* (Algiers: Editions Dahlab, 1990).

Bradbrook, M.C., *Barbara Bodichon, George Eliot and the Limits of Feminism* (Oxford: Hollywell Press, 1975).

Brahimi, Denise, *Voyageurs français du XVIIIe siècle en Barbarie* (Paris: Honoré Champion, 1976).

Brau, Paul, *Trois siècles de médecine coloniale française* (Paris: Vigot, 1931).

Bresc, Henri and Christiane Veauvy (eds), *Mutations d'identités en Méditerranée: Moyen âge et époque contemporaine* (Paris: Editions Bouchène, 2000).

Brett, Michael and Elizabeth Fentress, *The Berbers* (Oxford: Blackwell, 1996).

de Broisard, M. le général, *Quatre-vingt-deux jours de commandement de la province d'Oran* (Perpignan: Jean-Baptiste Alzine, 1838).

Brunton, Deborah, *Medicine Transformed: Health, Disease and Society in Europe 1800– 1930* (Manchester: Manchester University Press, 2004).

Bugeaud, M., *De l'établissement de légions de colons militaires dans les possessions françaises du nord de l'afrique* (Paris: Firmin Didot, 1838).

Bugeaud de la Piconnerie, Thomas Robert, *De la Colonisation de l'Algérie* (Paris: A. Guyot, 1847).

Burke III, Edmund and David N. Yaghoubian (eds), *Struggle and Survival in the Modern Middle East*, 2nd edn (Berkeley: University of California Press, 2006).

Burleigh, Michael. *Death and Deliverance: 'Euthanasia' in Germany 1900–1945* (Cambridge: Cambridge University Press, 1994).

Burton, Hester, *Barbara Bodichon 1827–1891* (London: John Murray, 1949).

Cabrol, Dr, *De l'Algérie sous le rapport de l'hygiène et de la colonisation* (Paris: Challamel Ainé, 1863).

Camau, Michel, Hédi Zaiem and Hajer Bahri, *État de santé: besoin medical et enjeux Politiques en Tunisie* (Paris: ECNRS, 1990).

Campbell, Alastair, Grant Gillett and Gareth Jones (eds), *Medical Ethics* (Oxford: Oxford University Press, 1997).

Campbell, Alastair, ' "My Country Tis of Thee" – The Myopia of American Bioethics', *Medicine, Health Care and Philosophy*, 3 (2000), 195–98.

Carette, M.M. and Auguste Warnier, *Description et division de l'Algérie* (Paris: Hachette, 1847).

Carr, David, Thomas R. Flynn and Rudolf A. Makkreel (eds), *The Ethics of History* (Evanston: Northwestern University Press, 2004).

de Castellane, The Count P., *Military Life in Algeria*, 2 vols (London: Hurst and Blackett, 1853).

Caze, F., *Notice sur Alger* (Paris: Félix Locquin, 1831).

Charbonneau, Pierre, 'La politique hospitalière pendant le Protectorat', *Histoire des sciences médicales*, 28 (1994), 167–169.

Charby, Jacques, *Les Porteurs d'espoir: Les réseaux de soutien au FLN pendant la guerre d'Algérie – les acteurs parlent* (Paris: La Découverte, 2004).

Charlton, D.G., *Secular Religions in France 1815–1870* (Oxford: Oxford University Press, 1963).

Chase, A., *The Legacy of Malthus: The Social Cost of the New Scientific Racism* (New York: Knopf, 1977).

Chatinière, Paul, *Dans le Grand Atlas Marocain* (Paris: Plon, 1910).

Cherry, Deborah, *Beyond the Frame: Feminism and Visual Culture, Britain 1850–1900* (London: Routledge, 2000).

Du Cheyron, Le Commandant, *Bordj-Bou-Arréridj pendant l'insurrection de 1871 en Algérie: Journal d'un officier* (Paris: Henri Plon, 1873).

Childress, James F., *Practical Reasoning in Bioethics* (Bloomington: Indiana University Press, 1997).

Christelow, Allan, *Muslim Law Courts and the French Colonial State in Algeria* (Princeton: Princeton University Press, 1985).

Christian, P., *Souvenirs du Maréchal Bugeaud de l'Algérie et du Maroc* (Paris: Alexandre Cadot, 1845).

Citati, Gaétan, *Des propriétés et des propriétaires d'Alger* (Toulon: Eugène Aurel, 1840).

Citati, Gaétan, *Essai sur la nécessité de créer une vice-royauté en Algérie* (Marseille: Carnaud, 1847).

Claisse-Dauchy, Renée, *Médecine traditionnelle du Maghreb: ritual d'envoûtement et de guérison au Maroc* (Paris: L'Harmattan, 1996).

Clancy-Smith, Julia (ed.), *North Africa, Islam and the Mediterranean World* (London: Routledge, 2001).

Clancy-Smith, Julia, 'The Shaykh and his Daughter: Coping in Colonial Algeria', in Burke III, Edmund and David N. Yaghoubian (eds), *Struggle and Survival in the Modern Middle East*, 2nd edn (Berkeley: University of California Press, 2006), 119–136.

Clancy-Smith, Julia and Frances Gouda (eds), *Domesticating the Empire: Race, Gender and Family Life in French and Dutch Colonialism* (Charlottesville: University Press of Virginia, 1998).

Clarac, Charles, *Conséquences de l'ignorance et de la mainpulations des français sur l'Islam* (Nimes: C. Lacour, 1994).

Clauzel, Maréchal, *Nouvelles observations de M. le Maréchal Clauzel sur la colonisation d'Alger addressées à M. le Maréchal, ministre de la guerre, président du conseil* (Paris: Imprimerie Selligue, 1833).

Cluseret, G., *Mémoires du général Cluseret* (Paris: Calmann Lévy, 1887).

Cœur, Pierre, *Les derniers de leur race* (Paris: Paul Ollendorff, 1885).

Cœur, Pierre, *Un drame à Alger* (Paris: Calmann Lévy, 1887).

Cœur, Pierre, *L'Assimilation des indigènes musulmans* (Paris: A. Guédan, 1890).

Cohen, Jacques, *Les Israélites de l'Algérie et le décret Crémieux* (Paris: Arthur Rousseau, 1900).

Cohen, William B., *The French Encounter with Africans: White Responses to Blacks, 1530–1880* (Bloomington: Indiana University Press, 1980).

Coinze d'Altroff, M., *Introduction à un plan général d'administration civile et de colonisation agricole en Algérie* (Paris: Jules Frey, 1847).

Colombani, Jules, *La Ministère de la santé et de l'hygiène publiques au Maroc*, (Casablanca: Editions France-Marocaine, n.d.).

Les colons d'Alger à la France: domination générale, colonisation progressive, gouvernement civil (Marseille: Marius Olive, 1840).

Conklin, Alice L., *A Mission to Civilize: The Republican Idea of Empire in France and West Africa, 1895–1930* (Stanford: Stanford University Press, 1997).

Coomans, Casimir, *De Marseille à Gênes par la Corniche – En Algérie* (Brussels: Coomans, 1880).

Cooter, Roger, 'The Resistible Rise of Medical Ethics', *Social History of Medicine*, 8/2 (1995), 257–270.

Cooter, Roger, Mark Harrison and Steve Sturdy (eds), *War, Medicine and Modernity* (Stroud: Sutton, 1998).

Cooter, Roger, Mark Harrison and Steve Sturdy (eds), *Medicine and Modern Warfare* (Amsterdam: Rodopi, 1999).

Cooter, Roger and John Pickstone (eds), *Medicine in the Twentieth Century* (Amsterdam: Harwood, 2000).

Corbin, Alain, 'Douleurs, souffrances et misères du corp', in Alain Corbin, Jean-Jacques Courtine, Georges Vigarello (eds) *Histoire du Corps*, 2 vols, II (Paris: Éditions du Seuil, 2005), 215–273.

Corbin, Alain, 'La recontre des corps', in Alain Corbin, Jean-Jacques Courtine, Georges Vigarello (eds), *Histoire du Corps*, 2 vols, II (Paris: Éditions du Seuil, 2005), 149–214.

Corten, André and Marie-Blanche Tahon, *L'État nourricier: prolétariat et population – Mexique/Algérie* (Paris: L'Harmattan, 1988).

di Costanzo, Jean-Maurice, *Allemands et Suisses en Algérie: 1830–1918* (Nice: Éditions Jacques Gandini, 2001).

Crosby, Alfred W., *Ecological Imperialism: The Biological Expansion of Europe, 900–1900* (Cambridge: Cambridge University Press, 1986).

Cunningham, Andrew and Bridie Andrews (eds), *Western Medicine as Contested Knowledge* (Manchester: Manchester University Press, 1997).

Cunningham, Andrew and Perry Williams (eds), *The Laboratory Revolution in Medicine* (Cambridge: Cambridge University Press, 2002).

Curtin, Philip D., 'Disease and Imperialism', in Arnold, David (ed.), *Warm Climates and Western Medicine: The Emergence of Tropical Medicine 1500–1800* (Amsterdam: Rodopi, 1996), 99–107.

Curtin, Philip D., *Disease and Empire: The Health of European Troops in the Conquest of Africa* (Cambridge: Cambridge University Press, 1998).

Davies, David Cornelson, *Frantz Fanon, Colonialism and Algeria: The Historical Formation of a Radical Discourse*, University of Brighton PhD thesis, 2004.

Davis, Diana K., *Resurrecting the Granary of Rome: Environmental History and the French Colonial Expansion in North Africa* (Ohio: Ohio University Press, 2007).

Davis, Mike, *Late Victorian Holocausts: El Niño Famines and the Making of the Third World* (London: Verso, 2001).

Decaux, Alain and Nicole Granier, *Abd el-Kader et l'Algérie au XIXe siècle dans les collections du musée Condé à Chantilly* (Paris: Somogy, 2003).

Delavigne, Paul, Auguste Warnier, O.MacCarthy, U. Ranc, J.A. Serpolet *Chemin de fer de l'Algérie par la ligne centrale du Tell* (Algiers: Beau, 1854).

Delumeau, Jean and Yves Lequin (eds), *Les Malheurs du temps: histoire des fléaux et des calamités en France* (Paris: Larousse, 1987).

Desrouelles, Maurice and Henri Bersot, 'Care of the insane since the C19', in *History of Psychiatry* special issue 549/561 (1996) – orig. pub. 1939.

du Dézen, Delaunay, *Manuel du futur colon en Algérie* (Paris: Augustin Challamel, 1892).

dell'Oro, Roberto and Corrado Viafora (eds), *Bioethics: A History – International Perspectives* (San Francisco: ISP, 1996).

Delpech de Saint-Guilhem, E., *Addresse de la délégation de l'Algérie aux chambres* (Paris: De Rignoux, 1847).

Delpech de Saint-Guilhem, E. et al., *Mémoire au roi et aux chambres par les colons de l'Algérie* (Paris: De Rignoux, 1847).

Derrécagaix, Général, *Le Maréchal Pélissier, Duc de Malakoff* (Paris: Librairie Militaire R. Chapelot, 1911).

Desjobert, A., *La Question d'Alger: Politique, Colonisation, Commerce* (Paris: A. Dufart, 1837).

Desjobert, A., *L'Algérie en 1838* (Paris: P. Dufart, 1838).

Desjobert, A., *L'Algérie en 1844* (Paris: Guillaumin, 1844).

Desjobert, A., *L'Algérie en 1846* (Paris: Guillaumin, 1846).

Desjobert, A., *État sanitaire de l'armée* (Paris: Panckoucke, 1848).

Dias, Jill, 'Famine and Disease in the History of Angola, c.1830–1930', *Journal of African History*, 22 (1981), 349–378.

Didier, Henry, *L'Algérie et le décret du 24 novembre* (Paris: E. Causin, 1861).

Dine, Philip, 'Shaping the Colonial Body: Sport and Society in Algeria, 1870–1962', in Lorcin, Patricia M.E. (ed.), *Algeria and France 1800–2000: Identity, Memory, Nostalgia* (Syracuse: Syracuse University Press, 2006), 33–48.

de Dombasle, C.J.A. Mathieu, *De l'avenir de l'Algérie* (Paris: Dufart, 1838).

Doury, Paul, *Henry Foley: apôtre du Sahara et de la médecine* (Helette: Jean Curutchet, 1998).

Drohojowska, La Comtesse, *Histoire de l'Algérie racontée à la jeunesse* (Paris: A. Allouard, 1848).

Ducrot, A., *La Vérité sur l'Algérie* (Paris: E. Dentu, 1871).

Dunwoodie, Peter, *Writing French Algeria* (Oxford: Oxford University Press, 1999).

Dunwoodie, Peter, 'Assimilation, Cultural Identity, and Permissible Deviance in Francophone Algerian Writing of the Interwar Years', in Lorcin, Patricia M.E. (ed.), *Algeria and France 1800–2000: Identity, Memory, Nostalgia* (Syracuse: Syracuse University Press, 2006)., 63–83.

Dupin, Le Baron Charles, *Situation comparée des colonies françaises et des colonies anglaises* (Paris: Firmin Didot, 1844).

Dupré, Auguste, *Exposé d'un projet d'ensemble de colonisation* (Algiers: A. Bouyer, 1863).

Dupré, Auguste, *Lettres sur l'Algérie* (Bordeaux: G. Gounouilhou, 1870).

Durand-Evrard, Françoise and Lucienne Martini (eds), *Archives d'Algérie 1830–1960* (Paris: Hazan, 2003).

Duval, Jules and Dr Auguste Warnier, *Bureaux Arabes et Colons: réponse au constitutionnel pour faire suite aux lettres de M. Rouher* (Paris: Challamel Ainé, 1869).

Duvernois, Clément, *La Réorganisation de l'Algérie: Lettre à S.A.I. le Prince Napoléon, chargé du ministère de l'Algérie et des colonies*, 2nd edn (Algiers: Dubos Frères, 1858).

Duvernois, Clément, *La Réaction, deuxième lettre* (Algiers: Dubos Frères, 1859).

Duvernois, Clément, *La Lieutenance de l'Empire*, 2nd edn (Algiers: Dubos Frères, 1859).

Duvernois, Clément, *L'Esprit et la lettre: lettre à M. Guillemard, Procureur Général à Alger* (Algiers: Dubos Frères, 1860).

Duvernois, Clément, *La Liberté de discussion: lettre à M. Levert, préfet d'Alger* (Algiers: Dubos Frères, 1860).

Duvernois, Clément, *L'Algérie pittoresque* (Paris: J. Rouvier, 1863).

Echenberg, Myron, *Black Death, White Medicine: Bubonic Plague and the Politics of Public Health in Colonial Senegal, 1914–1945* (Oxford: James Currey, 2002).

Eckenwiler, Lisa A. and Felicia G. Cohn (eds), *The Ethics of Bioethics: Mapping the Moral Landscape* (Baltimore: The Johns Hopkins University Press, 2007).

Ellis, Jack D., *The Physician Legislators of France: Medicine and Politics in the Early Third Republic 1870–1914* (Cambridge: Cambridge University Press, 1990).

El Tayeb, Salah El Din El Zein, *The National Ideology of the Radical Algerians and the Formation of the FLN 1924–1954* (Durham: Centre for Middle Eastern and Islamic Studies, 1987).

Emerit, Marcel (ed.), *La Révolution de 1848 en Algérie* (Paris: Éditions Larose, 1949).

Emerit, Marcel, *L'Algérie à l'époque d'Abd-el-Kader* (Paris: Éditions Bouchène, 2002 [1951]).

Enfantin, Barthélemy Prosper, *Colonisation de l'Algérie* (Paris: P. Bertrand, 1843).

Ernst, Waltraud and B. Harris (eds), *Race, Science and Medicine, 1700–1960* (London: Routledge, 1999).

Esquer, G. and P. Boyer, 'Bugeaud en 1840', *Revue Africaine*, 462/63 (1960), 57–98.

Esquisse Historique et Médicale de l'expédition d'Alger en 1830, par un officier de santé attaché au quartier-général de l'armée d'Afrique (Paris: Firmin Didot, 1831).

Evleth, Donna, 'Vichy France and the Continuity of Medical Nationalism', *Social History of Medicine*, 8/1 (1995), 95–116.

Explications du Maréchal Clauzel (Paris: Ambroise Dupont, 1837).

Exploration scientifique de l'Algérie pendant les années 1840, 1841, 1842, par ordre du gouvernement et avec le concours d'une commission académique, vol I – sciences médicales (Paris: Imprimerie Royal, 1847).

Fanon, Frantz, *L'An V de la revolution algérienne*, 2nd edn (Paris: François Maspéro, 1960).

Fanon, Frantz, *Les damnés de la terre* (Paris: François Maspéro, 1961).

Fanon, Frantz, *Pour la revolution africaine: écrits politiques* (Paris: François Maspéro, 1964).

Fanon, Frantz, *Peau Noire, Masques Blancs*, 2nd edn (Paris: Éditions du Seuil, 1965).

Fanon, Frantz, *Toward the African Revolution* (London: Monthly Review Press, 1967).

Fanon, Frantz, *Dying Colonialism* (London: Grove, 1988).

Farley, John, *Bilharzia: A History of Imperial Tropical Medicine* (Cambridge: Cambridge University Press, 1991).

Faure, Olivier, 'The Social History of Health in France: A Survey of Recent Developments', *Social History of Medicine*, 3/3 (1990), 437–451.

Faure, Olivier, *Les Français et leur médecine au XIXe siècle* (Paris: Éditions Belin, 1993).

Faure, Olivier and Dominique Dessertine, *La Maladie entre libéralisme et solidarities (1850–1940)* (Paris: Mutualité française, 1994).

Fechner, Elisabeth, *Là-bas la France: Souvenirs d'une Algérie heureuse* (Paris: Calmann Lévy, 2003).

Ferro, Marc, 'La conquête de l'Algérie', in Ferro, Marc (ed.), *Le livre noir du colonialisme: XVIe – XXIe siècle: de l'extermination à la repentance* (Paris: Robert Laffont, 2003), 490–505.

Féry, Raymond, *Médecin chez les berbères* (Versailles: Éditions de l'Atlanthrope, 1986).

Filleul de Pétigny, Clara, *L'Algérie* (Clermont: Thibaud-Landriot, 1851).

Flandin, Jean-Baptiste, *Régence d'Alger: Peut-on la coloniser? Comment?* (Paris: Féret, 1833).

Flandin, Jean-Baptiste, *De la Régence d'Alger. Solution de ces questions: doit-on conserver cette régence? Peut-on la coloniser? Comment?* (Paris: Anselin, 1834).

Flandin, Jean-Baptiste, *Notice sur la prise de possessions des trésors de la régence d'Alger* (Paris, 1848).

Fleury, Georges, *Comment l'Algérie devint française 1830–1848* (Paris: Perrin, 2004).

Flint, Karen, 'Competition, Race, and Professionalization: African Healers and White Medical Practitioners in Natal, South Africa in the Early Twentieth Century', *Social History of Medicine*, 14/2 (2001), 199–221.

Fogarty, Richard and Michael Osborne, 'Constructions and Functions of Race in French Military Medicine, 1830–1920', in Peabody, Sue and Tyler Stovall (eds), *The Color of Liberty: Histories of Race in France* (Durham: Duke University Press, 2003), 206–36.

Forissier, Régis, 'L'assistance médicale gratuite apporté par l'armée aux populations nécessiteuses d'Algérie entre 1954 et 1962', *Revue historique de l'armée*, 3 (1995), 29–44.

Foucault, Michel, *Naissance de la clinique: Une archéologie du regard médical* (Paris: PUF, 1963).

Foucault, Michel, *Surveillir et punir* (Paris: Gallimard, 1975).

La France: doit-elle conserver Alger? par un auditeur au conseil d'état (Paris: Bethuné et Plon, 1835).

de Franclieu, M. le Comte, *Encore l'Algérie devant les chambres* (Paris: Bureau de la revue algérienne, 1847).

Frémeaux, Jacques, *La France et l'Islam depuis 1789* (Paris: PUF, 1991).

Frémeaux, Jacques, *La France et l'Algérie en guerre 1830–1870, 1954–1962*, (Paris: Economica, 2002).

Fumeron d'Ardeuil, M., *Observations sur la situation et l'avenir de nos possessions d'afrique* (Paris: Éditions Belin, 1839).

Gallagher, Nancy Elizabeth, *Medicine and Power in Tunisia, 1780–1900* (Cambridge: Cambridge University Press, 1983).

Gallagher, Nancy Elizabeth, 'Medicine: Contemporary Practices', *The Oxford Encyclopedia of the Modern Islamic World*, vol. III (1995), 89–91.

Gallois, William, 'Todorov's Gift of Ethics to History', *Canadian Review of Comparative Literature*, 31/2 (2004), 195–210.

Garber, Marjorie, Beatrice Hanssen and Rebecca Walkowitz (eds), *The Turn to Ethics* (London: Routledge, 2000).

Gendzier, Irene L., *Frantz Fanon: A Critical Study* (London: Wildwood House, 1973).

Ghoti, Mohamed, *Histoire de la médecine au Maroc: le 20e siècle, 1896–1994* (Casablanca: Imprimerie Idéale, 1995).

Gill, Hélène, *The Language of French Orientalist Painting* (Lewiston: Edwin Mellen Press, 2003).

Gilman, Sander L., *Difference and Pathology: Stereotypes of Sexuality, Race and Madness* (Ithaca: Cornell University Press, 1998).

Glover, Jonathan, *Humanity: A Moral History of the Twentieth Century* (London: Jonathan Cape, 1999).

Goldstein, Jan, *Console and Classify: The French Psychiatric Profession in the Nineteenth Century* (Cambridge: Cambridge University Press, 1987).

Goodin, Robert, 'What is so special about our fellow countrymen?' *Ethics*, 98 (1988), 663–686.

Gordon, Lewis R., *Fanon and the Crisis of European Man: An Essay on Philosophy and the Human Sciences* (London: Routledge, 1995).

Gordon, Lewis R. (ed.), *Fanon: A Critical Reader* (Oxford: Blackwell, 1996).

Goubert, J.P., 'La médicalisation de la société française 1770–1830', *Réflexions historiques/ Historical Reflections*, 9 (Ontario: University of Waterloo Press, 1982).

de la Gournerie, Eugène, *La Béarnaise: Épisode des geurres d'Afrique* (Paris: Paulin, 1834).

Graebner Seth, ' "Unknown and Unloved" The Politics of French Ignorance in Algeria, 1860–1930', in Lorcin, Patricia M.E. (ed.), *Algeria and France 1800–2000: Identity, Memory, Nostalgia* (Syracuse: Syracuse University Press, 2006), 49–62.

Grandin, Léonce, *Le maréchal Pélissier, Duc de Malakoff* (Abbeville: C. Paillart, 1901).

de Grieu, René, *Le Duc d'Aumale et l'Algérie* (Paris: Blériot et Gautier, 1884).

Grosvenor, Lord R., *Extracts from the Journal of Lord R. Grosvenor: Being an Account of His Visit to the Barbary Regencies in the Spring of 1830* (Chester: G. Harding, 1830).

Grove, Richard H., *Green Imperialism: Colonial Expansion, Tropical Island Edens and the Origins of Environmentalism, 1600–1860* (Cambridge: Cambridge University Press, 1995).

Gsell, G., G. Marçais and G. Yver, *Histoire d'Algérie* (Paris: Boivin, 1929).

Guernier, Eugène (ed.), *Encyclopédie de l'empire français: Algérie et Sahara* (Paris: Encyclopédie de l'empire français, 1946).

Guiard, Emile, *La trépanation cranienne chez les néolithiques et chez les primitifs modernes* (Paris: Masson, 1930).

Guillaumet, Gustave, *Catalogue des tableaux, dessins, pastels et aquarelles provenant de l'atelier G. Guillaumet* (Paris: Imprimerie de l'art, 1888).

Guillaumet, Gustave, *Tableaux algériens* (Paris: Plon, 1888).

Gutas, Dimitri, *Greek Thought, Arabic Culture: The Graeco-Arabic Translation Movement in Baghdad and Early Abbasid Society* (London: Routledge, 1998).

Guy, Alfred, *Des Famines périodiques en Algérie et d'un moyen d'y porter remède* (Paris: Auguste Challamel, 1893).

Hadjeres, Sadek, *Culture, indépendance et révolution en Algérie* (Paris: Temps Actuel, 1981).

Hanafy, A. Youssef and Salah A. Fadl, 'Frantz Fanon and Political Psychiatry', *History of Psychiatry*, special issue (1996), 525–532.

Harbi, Mohammed, *Le FLN: Mirage et Réalité* (Paris: Editions J.A., 1980).

Harbi, Mohammed and Gilbert Meynier, *Le FLN: Documents et Histoire 1954–1962* (Paris: Fayard, 2004).

Haroun, Ali, *La 7e Wilaya: La Guerre du FLN en France 1954–1962* (Paris: Editions du Seuil, 1986).

Harris, George W., *'The' Practical Guide to Algiers* (London: George Philip, 1890).

Harrison, Mark, 'Medicine and the Management of Modern Warfare: An Introduction', in Cooter, Roger, Mark Harrison and Steve Sturdy (eds), *Medicine and Modern Warfare* (Amsterdam: Rodopi, 1999), 1–27.

Harrison, Mark, *Medicine and Victory: British Military Medicine in the Second World War* (Oxford: Oxford University Press, 2004).

Haynes, Douglas M., *Imperial Medicine: Patrick Manson and the Conquest of Tropical Disease* (Pennsylvania: University of Pennsylvania Press, 2001).

Hebey, Pierre, *Alger 1898: La grande vague antijuive* (Paris: NiL, 1996).

Heifetz, Milton D., *Ethics in Medicine* (Amherst: Prometheus, 1996).

Henissart, Paul, *Wolves in the City: The Death of French Algeria* (London: Rupert Hart-Davis, 1970).

Herbert, Lady, *A Search after Sunshine: Algeria in 1871* (London: Richard Bentley, 1872).

Herstein, Sheila R., *A Mid-Victorian Feminist: Barbara Leigh Smith Bodichon* (Yale: Yale University Press, 1985).

Hilton-Simpson, M.W., *Arab Medicine and Surgery: A Study of the Healing Art in Algeria* (London: Oxford University Press, 1922).

Hirsch, Pam, *Barbara Leigh Smith Bodichon 1827–1891: Feminist, Artist and Rebel* (London: Chatto & Windus, 1998).

Horne, Alistair, *A Savage War of Peace: Algeria 1954–62*, 3rd edn (London: Pan, 2002).

Hourani, G.F., *Reason and Tradition in Islamic Ethics* (Cambridge: Cambridge University Press, 1985).

Hugues, Albert, *La Nationalité française chez les musulmans de l'Algérie* (Paris: A. Chevalier-Marescq, 1899).

Hussain, Abdilahi Bulhan, *Frantz Fanon and the Psychology of Oppression* (New York: Plenum Press, 1985).

Hutchinson, Martha Crenshaw, *Revolutionary Terrorism: The FLN in Algeria, 1954–1962* (Stanford: Hoover Institution Press, 1978).

d'Ideville, H., *Le Maréchal Bugeaud d'après sa correspondance intime et des documents inédits, 1784–1849*, 3 vols (Paris: Firmin-Didot, 1881).

d'Ideville, H., *Les Petits côtés de l'histoire: notes intimes et documents inédits, 1870–1884* (Paris: Calmann Lévy, 1884).

d'Ideville, H., *Memoirs of Marshal Bugeaud: From his Private Correspondence and Original Documents, 1784–1849*, 2 vols (London: Hurst and Blackett, 1884).

Jackson, G.A., *Algiers: Being a Complete Picture of the Barbary States: Their Government, Laws, Religion, and Natural Productions* (London: R. Edwards, 1817).

Jackson, Henry F., *The FLN in Algeria: Party Development in a Revolutionary Society* (Westport: Greenwood Press, 1977).

Janssens, P.G., 'The Colonial Legacy: Health and Medicine in the Belgian Congo', *Tropical Doctor*, 11/3 (1981), 132–140.

Jobin, Georges, *La Vie et l'œuvre de Maillot (1804–1894)* (Paris: Marcel Vigné, 1931).

Joly, M., *Discours prononcé dans la discussion du projet de loi relatif à divers crédits extraordinaires pour le service de l'Algérie*, Chamber of Deputies, séance du 23 mai 1843.

Jordanova, Ludmilla, 'The Social Construction of Medical Knowledge', *Social History of Medicine*, 7/3 (1995), 361–381.

Jordi, Jean-Jacques and Jean-Louis Planche (eds), *Alger 1830–1939: Le modèle ambigu du Triomphe colonial* (Paris: Editions Autrement, 1999).

Jonsen, Albert R., *The Birth of Bioethics* (Oxford: Oxford University Press, 1998).

Jonsen, Albert R., *A Short History of Medical Ethics* (Oxford: Oxford University Press, 2000).

Julien, Charles-André, *Une pensée anticoloniale: positions 1914–1979* (Paris: Sindbad, 1979).

Jung, Eugène, *L'Islam et les Musulmans dans l'Afrique du Nord* (Paris: Jeune Parque, 1930).

Kahl, Oliver (ed.), *Ya'qūb ibn Ishāq al-Isrā'īlī's 'Treatise on the Errors of the Physicians of Damascus'* (Oxford: Oxford University Press, 2000).

Kahya, Esin and A. Demirhan Erdemir, *Medicine in the Ottoman Empire (And Other Scientific Developments)* (Istanbul: Nobel Medical, 1997).

Kamel Kateb, *La Statistique colonial en Algérie (1830–1962): Entre la reproduction du système métropolitain et les impératifs d'adaptation à la réalité algérienne*: http://www. insee.fr/fr/ffc/docs_ffc/cs112b.pdf

Karmi, Ghada, 'The colonisation of traditional Arabic medicine', in Porter, Roy (ed.) *Patients and Practitioners: Lay Perceptions of Medicine in Pre-industrial Medicine* (Cambridge: Cambridge University Press, 2003), 315–340.

Katz, Jonathan G., 'The 1907 Mauchamp Affair and the French Civilizing Mission in Morocco', in Clancy-Smith, Julia (ed.), *North Africa, Islam and the Mediterranean World* (London: Routledge, 2001), 143–166.

Keller, Richard, 'Madness and Colonization: Psychiatry in the British and French Empires, 1800–1962', *Journal of Social History* (2001), 295–326.

Kelsch, A. and P.L. Kiener, *Traité des maladies des pays chauds: région prétropicale* (Paris: J.B. Baillière, 1889).

Khan, Muhammad Salim, *Islamic Medicine* (London: Routledge, 1986).

Khodja, Hamdan, *Le Miroir: Aperçu historique et statistique sur la Régence d'Alger* (Paris: Sindbad, 1985).

Khoja, Sidy Hamdan-ben-Othman, *Aperçu historique et statistique sur la Régence d'Alger, intitulé en arabe Le Miroir* (Paris: Gœtschy, 1833).

Kirkup, John, 'Shawiya Berber Surgical Instruments in the Aurès Mountains, 1913–1922', Proceedings of the 36th International Congress on the History of Medicine, Tunis, 1986, 149–156.

Kob, Jacques L., *L'Algérie: un moyen pratique pour faire un pas en avant* (Paris: Sandoz and Fischbacher, 1880).

Kolb, Edmond, *Études sur l'Hygiène de l'Algérie*, Faculty of Medicine, University of Montpellier PhD thesis, 1859 (Montpellier: L. Cristin).

La Berge, Anne F., *Mission and Method: The Early Nineteenth-Century French Public Health Movement* (Cambridge: Cambridge University Press, 1992).

La Berge, Anne and Mordechai Feingold (eds), *French Medical Culture in the Nineteenth Century* (Amsterdam: Rodopi, 1994).

Labidi, Djamel, *Science et pouvoir en Algérie*, 2 vols (Algiers: OPU, 1993).

Lacey, Candida Anne (ed.), *Women's Source Library*, vol III: *Barbara Leigh Smith Bodichon and the Langham Place Group*, London: Routledge.

Laignel-Lavastine, Maxime and Raymond Molinéry, *French Medicine* (New York: Paul B. Hoeber, 1934).

Lamarche, Hippolyte, *L'Algérie, son influence sur les destinées de la France et de l'Europe* (Paris: Paulin, 1846).

Lamarque, Léo, *De la conquête et de la colonisation de l'Algérie* (Paris: Ancelin, 1841).

Lamarque, Léo, *Chemins de fer et canaux en Algérie* (Algiers: Dubos Frères, 1847).

Lapeyssonnie, Léon, *La Médecine Coloniale: Mythes et Réalités* (Paris: Seghers, 1988).

Lasry, Albert, *Histoire de la Pharmacie Indigène de l'Algérie et son Folklore* (Oran: Achour frères, 1937).

Latour, Bruno, *The Pasteurization of France* (Cambridge, Mass.: Harvard University Press, 1988).

Leblanc de Prébois, François, *Lettre sur l'Algérie à messieurs les membres de la chambre des députés* (Montpellier: Boehm, 1840).

Leblanc de Prébois, François, *Bilan du régime civil de l'Algérie à la fin de 1871* (Paris: E. Dentu, 1872).

Leclerc, Lucien, *Histoire de la Médecine Arabe – Exposé complet des traductions du grec – Les sciences en orient – Leur transmission à l'occident par leur traductions latines*, 2 vols (Paris: Ernest Leroux, 1876).

Le Cour Grandmaison, Olivier, *Coloniser. Exterminer: Sur la guerre et l'état colonial* (Paris: Fayard, 2005).

Le Coz, Raymond, *Les médecins nestoriens au moyen âge: Les maîtres des arabes* (Paris: L'Harmattan, 2004).

Ledentu, A., *Pourquoi l'Algérie a-t-elle jusqu'ici un fardeau pour la France?* (Paris: G.A. Dentu, 1845).

Léonard, Jacques, *Les Officiers de Santé de la Marine française de 1814 à 1835*, Université de Rennes PhD thesis, 1967.

Léonard, Jacques, *Les Medecins de l'ouest au XIXe siècle*, 3 vols (Paris: Honoré Champion, 1978).

Léonard, Jacques, *Archives du corps: la santé au XIXe siècle* (Rennes: Ouest-France, 1986).

Le Pays de Bourjolly, M. le lieutenant-général, *Considérations sur l'Algérie, ou les faits opposés aux théories* (Paris: Chez Tresse, 1846).

Le Pays de Bourjolly, M. le lieutenant général, *Projets sur l'Algérie* (Paris: Librairie Militaire de J. Dumaine, 1847).

Lepetit, Bernard, 'Missions scientifiques et expéditions militaires: remarques sur leurs modalités d'articulation' in Bourguet Marie-Noëlle, Bernard Lepetit, Daniel Nordman and Maroula Sinarellis (eds), *L'Invention scientifique de la Mediterranée: Egypte, Morée, Algérie* (Paris: Editions EHESS, 1998), 97–116.

Lesourd, Paul, *Les Pères Blancs du Cardinal Lavigerie* (Orléans: Bernard Grasset, 1935).

Létang, Le Maréchal-de-camp B., *Des moyens d'assurer la domination française en Algérie* (Paris: Anselin, 1840).

Lettre à un député sur l'administration civile en Algérie et les crédits demandés pour 1846 (Paris: Lange Levy, 1845).

Lettre d'un colon d'Alger à M. Blanqui (Marseille: Feissat and Demonchy, 1840).

Levallois, Michel, 'Ismayl Urbain: Eléments pour une biographie', in Morsy, Magali (ed.), *North Africa 1800–1900: A Survey from the Nile Valley to the Atlantic* (London: Longman, 1984), 53–82.

Levalllois, Anne, *Les écrits autobiographiques d'Ismayl Urbain: Homme de couleur, saint-slmonlen et musulman (1812–1884)* (Paris: Maisonneuve & Larose, 2005).

Lihoreau, Michel, *Une page de la conquête du Sahara: l'expédition Wimpffen à l'Oued Guir en 1870*, (Paris: L'Harmattan, 1996).

Limbourg, H., *Le Duc d'Aumale et sa deuxième campagne d'Afrique (février à septembre 1841)* (Paris: Édition spéciale de la revue hebdomadaire, 1915).

Loi (du 16 septembre 1871) sur le budget rectitatif de l'exercice 1871.

Longuenesse, Élisabeth (ed.), *Santé, médecine et société dans le monde arabe* (Paris: L' Harmattan, 1995).

Lorcin, Patricia M.E., *Imperial Identities: Stereotyping, Prejudice and Race in Colonial Algeria* (London: I.B. Tauris, 1995).

Lorcin, Patricia M.E., 'Imperialism, Colonial Identity and Race in Algeria, 1830–1870: The Role of the French Medical Corps', *Isis*, 90/4 (1999), 652–679.

Lorcin, Patricia M.E. (ed.), *Algeria and France 1800–2000: Identity, Memory, Nostalgia* (Syracuse: Syracuse University Press, 2006).

Lucas, Colin, *Peter Winch* (Teddington: Acumen, 1999).

Lyautey, Maréchal, Gaudefroy- Demombynes, Paul Boyer, Marcel Grant, Général Weygand, René Pinon, Jules Cambon, Augustin Bernard, comte de Saint-Aulaire, and Louis Massignon, *L'Islam et la politique contemporaine* (Paris: Félix Alcan, 1927).

Lyons, Marinyez, *The Colonial Disease: A Social History of Sleeping Sickness* (Cambridge: Cambridge University Press, 1992).

de la M, M., J., *Réflexions sur l'état actuel d'Alger* (Paris: Le Normant, 1836).

Macey, David, *Frantz Fanon: A Life* (London: Granta, 2000).

Macleod, Roy and Milton Lewis (eds), *Medicine and Empire: Perspectives on Western Medicine and the Experience of European Expansion* (London: Routledge, 1988).

Maehle, Andreas-Holger and Johanna Geyer-Kordesch (eds), *Historical and Philosophical Perspectives on Biomedical Ethics: From Paternalism to Autonomy?* (Aldershot: Ashgate, 2002).

Magraw, Roger, *France 1815–1914: The Bourgeois Century* (London: Fontana, 1983).

Malbot, Henri and R. Verneau, *Etude d'ethnographie algérienne: Les Chaouïas et la trépanation du crâne dans l'Aurès* (Paris: Masson, 1897).

Malfroy, M. B., Olivier, P. Bichet and J. Guiraud, *Histoire de Pontarlier* (Besançon: Cêtre, 1979).

Malherbe, Jean-Francois, 'Orientations and Tendencies of Bioethics in the French-Speaking World', in dell'Oro, Roberto and Corrado Viafora (eds), *Bioethics: A History – International Perspectives* (San Francisco: ISP, 1996), 119–154.

Malti-Douglas, Fedwa, *Medicines of the Soul: Female Bodies and Sacred Geographies in a Transnational Islam* (Berkeley: University of California Press, 2001).

Marbaud, P., *Coup d'œil sur l'Algérie pendant la crise de 1859–1860 et réflexions sur le décret relatif à la vente des terres domaniales* (Constantine: V. Guende, 1860).

Marcovich, Anne, 'French Colonial Medicine and Colonial rule: Algeria and Indochina', in Macleod, Roy and Milton Lewis (eds), *Medicine and Empire: Perspectives on Western Medicine and the Experience of European Expansion* (London: Routledge, 1988), 103–117.

Marguerite, Paul, *L'Algérie de nos jours* (Algiers: J. Gervais-Courtellemont, 1893).

de Marguerittes, Julie, *Italy and the War of 1859* (Philadelphia: George G. Evans, 1859).

Marks, Shula, 'What is Colonial about Colonial Medicine? And What Has Happened to Imperialism and Health?' *Social History of Medicine*, 10/2 (1997), 205–219.

Martin, Claude, *La Commune d'Alger (1870–1871)* (Paris: Éditions Heraklès, 1936).

Marty, Docteur A., *Islamisme: Mœurs médicales et privées – climatologie de l'Algérie – considérations sur l'atmosphère* (Monaco: Petit Monagesque, 1903).

Mattei, A., *Protestation contre les détracteurs du système administratif suivi actuellement en Algérie / Coup d'œuil sur les différentes dominations en Afrique* (Paris: E. Dentu, 1869).

Matthews, William, *The Flora of Algeria Considered in Relation to the Physical History of the Mediterranean Region and Supposed Submergence of the Sahara* (London: Edward Stanford, 1880).

de Maupassant, Guy, *Lettres d'Afrique (Algérie-Tunisie)* (Paris: La Boîte à Documents, 1997).

McCulloch, Jock, *Black Soul, White Artefact: Fanon's Clinical Psychology and Social Theory* (Cambridge: Cambridge University Press, 1983).

James McDougall, 'Martyrdom and Destiny: The Inscription and Imagination of Algerian History', in Makdisi, Ussama and Paul A. Silverstein (eds) *Memory and Violence in the Middle East and North Africa* (Bloomington: Indiana University Press, 2006), 50–72.

McIlrath, M.B., 'A History of Medical Ethics in the Non-Christian World before the Rise of Modern Medicine', University of Sydney PhD thesis, 1959.

Mclellan III, James E., 'Science, Medicine and French Colonialism in Old-Regime Haiti', in Meade, Teresa and Mark Walker (eds), *Science, Medicine and Cultural Imperialism* (London: Macmillan, 1991), 36–59.

Medieval Islamic Medicine: Ibn Ridwān's Treatise 'On the Prevention of Bodily Ills in Egypt', trans. and with an introduction by Michael W. Dols (Berkeley: University of California Press, 1984).

Mercier, Ernest, *L'Algérie en 1880* (Paris: Challamel Aîné, 1880).

Meyer, Jean, Jean Tarrade, Annie Rey-Goldzeiguer and Jacques Thobie, *Histoire de la France coloniale: Des origines à 1914* (Paris: Armand Colin, 1991).

Meynier, Gilbert, *L'Algérie révélée: la guerre de 1914–1918 et le premier quart du XXe siècle* (Geneva: Droz, 1981).

Micoleau-Sicault, Marie-Claire, *Les médecins français au Maroc (1912–1956): Combats en urgence* (Paris: L'Harmattan, 2000).

Midgley Clare (ed.), *Gender and imperialism* (Manchester: Manchester University Press, 1998).

Mitchell, B.R., *International Historical Statistics: Africa, Asia and Oceania*, 3rd edn (London: Palgrave Macmillan, 1998).

Mobacher: Journal Officiel, 5 December 1881, 3.

Le Moniteur Universel, 68, samedi 9 mars 1833.

Montagnon, Pierre, *La Conquête de l'Algérie: les Germes de la Discorde 1830–1871* (Paris: Pygmalion, 1986).

Morsy, Magali, *North Africa 1800–1900: A Survey from the Nile Valley to the Atlantic* (London: Longman, 1984).

Morsy, Magali (ed.) *Les Saint-Simoniens et l'orient: Vers la modernité* (Aix-en-Provence: Édisud, 1990).

Mortimer, Mimmi (ed.), *Maghrebian Mosaic* (Boulder: Lynne Reiner, 2001).

Moulin, Anne-Marie, 'Tropical without the Tropics: The Turning-point of Pastorian Medicine in North Africa', in Arnold, David (ed.), *Warm Climates and Western Medicine: The Emergence of Tropical Medicine 1500–1800* (Amsterdam: Rodopi, 1996), 160–180.

de Nadaillac Le Marquis, *Mémoire sur l'Algérie, présenté à la conférence d'Orsay*, 18 jan 1842.

Nahum, Henri, *La médecine française et les juifs 1930–1945* (Paris: L'Harmattan, 2006).

Nandy, Ashis, *The Intimate Enemy: Loss and Recovery of Self under Colonialism* (Delhi: Oxford University Press, 1983).

Nanji, Azim A., 'Medical Ethics and the Islamic Tradition', *Journal of Medicine and Philosophy*, 13 (1988), 257–275.

Napoléon III, *Lettre sur la politique de la France en Algérie adressée par l'empereur au maréchal de MacMahon* (Paris: Imprimerie Impériale, 1865).

Navarro, Vincente (ed.), *Imperialism, Health and Medicine* (London: Pluto Press, 1982).

Nordman, Daniel, 'L'exploration scientifique de l'Algérie: le terrain et le texte', in Bourguet, Marie-Noëlle, Bernard Lepetit, Daniel Nordman and Maroula Sinarellis (eds), *L'Invention scientifique de la Méditerranée: Egypte, Morée, Algérie* (Paris: Editions EHESS, 1998), 71–95.

Note sur la situation d'Algérie à la fin de janvier 1838; demandée par le général Bernard et remise le 3 février 1838 (Paris: Bourgogne et Martinet, 1839).

Notice sur le mode de gouvernement provisoirement établi dans le royaume d'Alger (Paros: A. Moreau, 1830).

Nouschi, André, *Enquête sur le niveau de vie des populations rurales constantinois de la conquête jusqu'en 1919* (Paris: PUF, 1961).

O'Connor, Kathleen Malone, 'Prophetic Medicine and Qur'anic Healing: Religious Healing Systems in Islam', in Winchester Brown, Joseph and Robin Barlow (eds), *Studies in Middle Eastern Health* (Ann Arbor: Centre for Middle Eastern and North African Studies, University of Michigan, 1999), 39–77.

Ong, Aihwa and Stephen J. Collier (eds), *Global Assemblages: Technology, Politics and Ethics as Anthropological Problems* (Oxford: Blackwell, 2005).

Osborne, Michael A., *Nature, the Exotic and the Science of French Colonialism* (Bloomington: Indiana University Press, 1994).

Osborne, Michael A., 'Resurrecting Hippocrates: Hygienic Sciences and the French Scientific Expeditions to Egypt, Morea and Algeria', in Arnold, David (ed.), *Warm Climates and Western Medicine: The Emergence of Tropical Medicine 1500–1800* (Amsterdam: Rodopi, 1996), 80–98.

Osborne, Michael A., 'French Military Epidemiology and the Limits of the Laboratory: the Case of Louis-Félix-Achille Kelsch', in Cunningham, Andrew and Perry Williams (eds), *The Laboratory Revolution in Medicine* (Cambridge: Cambridge University Press, 2002), 189–208.

Osborne, Michael A., 'Science and the French Empire', *Isis*, 96, 80–87.

Ossendowski, Ferdinand, *The Breath of the Desert: The Account of a Journey through Algeria and Tunisia* (NY: Kessinger, 2005), orig. pub. 1927.

Ott, Edm., *Etude sur la colonisation de l'Algérie et en particulier sur le département de Constantine* (Paris: Auguste Ghio, 1880).

Oufriha, Fatima-Zohra, *Système de santé et population en Algérie* (Algiers: Éditions ANEP, 2002).

Owen, Norman G. (ed.), *Death and Disease in South-East Asia: Explorations in Social, Medical and Demographic History* (Oxford: Oxford University Press, 1987).

Palladino, Paolo, 'And the Answer is…42', *Social History of Medicine*, 13/1 (2000), 147–151.

Parent-Duchatelet, A.J.B., *La Prostitution de Paris*, tome II (Paris: J.B. Baillière, 1857).

Parry, John T., 'Pain, Interrogation, and the Body: State Violence and the Law of Torture', in Parry, John T., (ed.), *Evil, Law and the State: Perspectives on State Power and Violence* (Amsterdam: Rodopi, 2006), 1–16.

Peabody, Sue and Tyler Stovall (eds), *The Color of Liberty: Histories of Race in France* (Durham: Duke University Press, 2003).

Péan, Pierre, *Main basse sur l'Alger* (Paris: Plon, 2004).

Pein, T., *Lettres familières sur l'Algérie, un petit royaume arabe* (Paris: Tanera, 1871).

Pelis, Kim, *Charles Nicolle: Pasteur's Imperial Missionary: Typhus and Tunisia* (Rochester: University of Rochester Press, 2006).

Pellissier de Raynaud, E., *Lettre à M. Desjobert sur la question d'Alger* (Algiers: Imprimerie du Gouvernement, 1837).

Pellissier de Raynaud, E., *Quelques mots sur la colonisation militaire en algérie* (Paris: Garnier Frères, 1847).

Pellissier de Reynaud, E., *Annales Algériennes*, 2nd edn, 3 vols (Paris: J. Dumaine, 1854).

Perkins, Kenneth J., *Qaids, Captains and Colons: French Military Administration in the Colonial Maghrib 1844–1934* (New York: Africana, 1981).

Perrot, A., *La Conquête d'Alger ou Relation de la Campagne d'Afrique* (Paris: H. Langlois, 1830).

Perrot, A.M., *Alger: Esquisse Topographique et Historique* (Paris: Ladvocat, 1830).

Pétition des colons d'alger, suivie des négocians de Marseille et des déliberations du conseil municipal et de la chambre de commerce de la même ville (Marseille: Feissat and Demonchy, 1834).

Peyronny, M., *Considérations politiques sur la colonie d'Alger* (Paris: G.A. Dentu, 1936).

Picot, J.B.C., *Colonisation de l'Algérie* (Paris: Moquet, 1848).

Plunkett, E., *The Past and Future of the British Navy* (London: Longman, 1846).

Poiriel, M., *De la déportation et de la colonisation de l'Algérie* (Paris: Hingray, 1844).

Porch, Douglas, *The Conquest of the Sahara* (Oxford: Oxford University Press, 1986).

Porter, Roy (ed.), *Patients and Practitioners: Lay Perceptions of Medicine in Pre-industrial Medicine* (Cambridge: Cambridge University Press, 2003).

Préaux-Locré, Le Colonel d'artillerie, *Mémoire sur l'Algérie* (Paris: Garnier Frères, 1846).

Premare de, A.L., *Ethique muslumane et relations sociales dans la famille maghrébine*, Université d'Aix-Marseille I, Mémoire de maitrise en psychologie, 1973.

Presbey, Gail M., 'Fanon on the role of violence in liberation: a comparison with Gandhi and Mandela', in Gordon, Lewis R. (ed.), *Fanon: A Critical Reader* (Oxford: Blackwell, 1996), 283–296.

Price, Richard, *Alabi's World* (Baltimore: The Johns Hopkins University Press, 1990).

Prochaska, David, *Making Algeria French: Colonialism in Bône 1870–1920* (Cambridge: Cambridge University Press, 1990).

Prochaska, David, 'The return of the repressed: war, trauma, memory in Algeria and beyond', in Lorcin, Patricia M.E. (ed.), *Algeria and France 1800–2000: Identity, Memory, Nostalgia* (Syracuse: Syracuse University Press, 2006). 257–276.

La Province de Constantine en 1839 et 1840 (Paris: Félix Locquin, 1843).

Puéjac, Mlle A., *L'Hygiène de la première enfance en Algérie* (Algiers: A. Bouyer, 1863).

Quelques mots addressés à la grande commission d'Alger, par M. Cappé, croix de juillet et avocat, député à Paris de cette colonie, au sujet de sa mission, et dédiés à la chambre des députés, à la chambre des pairs et au conseil d'état au nom des colons et des indigènes d'Alger (Paris: Gœtschy, 1834).

Quelques mots sur le trésor d'Alger (Paris: Dondey-Dupré, 1830).

Rahman, F., *Health and Medicine in the Islamic Tradition* (New York: Crossroad, 1987).

Rambaud, Pierre, *Le massacre de Crève-Cœur en Algérie* (Mascara: A. Albrecht, 1887).

Ramsey, Matthew, *Professional and Popular Medicine in France 1770–1830: The Social World of Medical Practice* (Cambridge: Cambridge University Press, 1988).

Rapport de M. Vatout surles crédits extraordinaires pour l'Algérie, séance du 13 mai 1843.

Rasteil, Maxime, *Un centre de colonisation en Algérie* (Bône: L. Rombi & Rasteil, 1895).

Raynard, J., *Restauration des forêts et des paturages du sud de l'Algérie (province d'Alger)* (Algiers: Adolphe Jourdan, 1880).

Razanajao, C., J. Postel, and D.F. Allen, 'The Life and Psychiatric Work of FF', in *History of Psychiatry*, special issue (1996), 499–524.

Read, James R., *The Study and Practice of Medicine by Women* (London: Chadwick, 1879).

Reclus, Onésime, *France, Algérie et Colonies* (Paris: Hachette, 1880).

Réfutation de l'ouvrage de Sidy Hamdan ben Othman Khoja intitulé Aperçu historique et statistique sur la régence d'Alger (Paris: Éverat, 1834).

Reggui, Marcel, *Les massacres de Guelma: Algérie, mai 1945; une enquête inédite sur la furie des milices coloniales* (Paris: La Découverte, 2006).

Reiser, Stanley Joel, Dick, Arthur J. and Curran, William J. (eds), *Ethics in Medicine: Historical Perspectives and Contemporary Concerns* (Cambridge, Mass.: The MIT Press, 1977).

Reland, Hadrien, *Institutions du droit mahométan relatives à la guerre sainte*, trans. Ch. Solvet (Algiers: Imprimerie du Gouvernement, 1838).

Renard, Emile, *De l'invasion des députés dans l'administration et particulièrement dans la nomination aux emplois publics* (Paris: Paul Dupont, 1844).

Renauld, Georges, *Adolphe Crémieux: Homme d'état français, juif et franc-maçon – le combat pour la république* (Paris: Detrad, 2002).

Renault, Eugène, *Première lettre à M. Passy, député, rapporteur du budget du ministre de la guerre, pour l'année 1836* (Paris: Panseron-Pinard, avril 1835).

Renault, François, *Cardinal Lavigerie: Churchman, Prophet and Missionary* (London: Athlone Press, 1992).

Reuillard, Michel, *Les Saint-Simoniens et la tentation coloniale: Les explorations africaines et le gouvernement néo-calédonien de Charles Guillain (1808–1875)* (Paris: Editions L'Harmattan, 1995).

Revue de l'Occident Musulman et de la Méditerranée, special issue on 'le Maghreb dans l'imaginaire français: la colonie, le désert, l'exil' (Aix-en-Provence: Edisud, 1985).

Rey-Goldzeiguer, Annie, *Le Royaume arabe: la politique algérienne de Napoléon III, 1861–1870* (Algiers: SNED, 1977).

Reynolds, E.A., *Mediterranean Winter Resorts: A Practical Handbook to the Principal Health and Pleasure Resorts on the Shores of the Mediterranean*, 2nd edn (London: Edward Stanford, 1890).

Richard, Charles, *Les Mystères du peuple arabe* (Paris: Challamel Ainé, 1860).

Rinn, L. 'Deux chansons Kabyles sur l'insurrection de 1871', *Revue Africaine*, journal publié par la Société historique d'Alger (1883).

Rinn, Louis, *Histoire de l'Insurrection de 1871 en Algérie* (Algiers: Librairie Adolphe Jourdan, 1891).

Rioux, Georges, *Dessin et Structure Mentale: Contribution à l'étude psycho-sociale des milieux nord-africains* (Paris: PUF, 1951).

Rispler-Chaim, Vardit, *Islamic Medical Ethics in the Twentieth Century* (Leiden: E.J. Brill, 1993).

Rivet, Daniel, *Hygiénisme colonial et médicalisation de la société marocaine au temps du Protectorat français (1912–56)* (Paris: L'Harmattan, 1995).

Roberts, Stephen H., *The History of French Colonial Policy 1870–1925* (London: Frank Cass, 1963).

Robin, Marie-Monique *Escadrons de la mort: l'école française* (Paris: La Découverte, 2004).

Rogniat, Le Général, *De la colonisation en Algérie et des fortifications propres à garantir les colons des invasions des tribus africains* (Paris: Gaultier-Laguionie, 1840).

Rothfield, Lawrence, *Vital Signs: Medical Realism in Nineteenth-Century Fiction* (Princeton: Princeton University Press, 1992).

Roussel, Napoléon, *Mon voyage en Algérie raconté à mes enfans* (Paris: Risler, 1840).

Rousso, Henry, *The Vichy Syndrome: History and Memory in France since 1944* (Cambridge, Mass.: Harvard University Press, 1991).

Rozet, M., *Relation de la Guerre d'Afrique pendant les années 1830 et 1831* (Paris: Firmin Didot, 1832).

Rozet, M., *Voyage dans la régence d'Alger ou description du pays occupé par l'armée française en Afrique*, 3 vols (Paris: Arthus Bertrand, 1833).

Rozet, C.A. and Antoine E.H. Carrette, *Algérie* (Paris: Firmin Didot, 1850).

de Rumigny, Le Comte, *Essai sur la province d'Alger sur les expéditions faites dans ce pays jusqu'à ce jour, et sur les moyens de les rendre les plus fructueuses suivi d'un essai sur l'influence de la vapeurs dans les guerres de terre et de mer* (Paris: Bourgogne et Martinet, 1841).

Russell, H., *Cardinal Lavigerie: Napoleon of Africa* (Dublin: Irish Messenger Office, 1945).

Russell-Crocker, Walter, *On Governing Colonies: Being an Outline of the Real Issues and a Comparison of the British, French and Belgian Approaches to Them* (London: George Allen & Unwin, 1947).

Sabatault, *De la nécessité d'établir un impot sur les grains importés de l'étranger: lettre à tous les amis de l'Algérie et de la France* (Algiers: Besancenez, 1845).

Said, Edward, *Orientalism* (London: Penguin, 1979).

Saidouni, Nacereddine, *L'Algerois Rural a la fin de l'époque Ottomane (1791–1830)* (Beirut: Dar al-gharb al-Islami, 2001).

Saint-Hypolite, *De l'Algérie: Moyens d'affermir nos possessions en 1840* (Paris: Bourgogne et Martinet, 1840).

Saint-Hypolite, *Notes sur le théâtre des opérations militaires dans le centre d'Algérie* (Paris: Bourgogne et Martinet, 1840).

Salva, Adolphe, *Quelques considérations hygiéniques sur Alger et ses habitans,,* Faculty of Medicine at Montpellier PhD thesis, 1832 (Montpellier: Auguste Ricard).

de Sarrauton, Henri, *La Question algérienne* (Oran: Paul Perrier, 1891).

Sartor, J., *Projet de réformes politiques et administratives de l'Algérie présenté à Mm. les membres de la commission chargée, par décret impérial, d'élaborer un projet de constitution pour l'Algérie* (Paris: Challamel Ainé, 1869).

Savage-Smith, Emilie, 'The Practice of Surgery in Islamic Lands: Myth and Reality', *Social History of Medicine*, 13/2 (2000), 307–321.

Savary, M., *Algérie: Nouveau projet d'occupation restreinte* (Paris: Anselin, 1840).

Schacht, M.J. and C.E. Bosworth, *The Legacy of Islam*, 2nd edn (Oxford: Oxford University Press, 1979).

Schleiner, Winfred, *Medical Ethics in the Renaissance* (Washington: Georgetown University Press, 1995).

Schreier, Joshua S., ' "They Swore upon the Tombs never To Make Peace with Us": Algerian Jews and French colonialism 1845–1848', in Lorcin, Patricia M.E. (ed.), *Algeria and France 1800–2000: Identity, Memory, Nostalgia* (Syracuse: Syracuse University Press, 2006). 101–116.

Scorer, Gordon and Wing, Antony (eds), *Decision Making in Medicine: the Practice if its Ethics* (London: Edward Arnold, 1979).

J. de S.-D., 'Quelques considérations sur le projet d'expulser les Ottomans de l'Europe et de les remplacer par une population gréco-slave', *Revue de l'orient: Bulletin de la société orientale*, 6 (1845), 286–293.

Sellam, Sadek, *L'Islam et les musulmans en France: perceptions, craintes et réalités* (Paris: Tougui, 1987).

Sergent, Edmond, *Les Travaux scientifiques de l'Institut Pasteur en Algérie de 1900 à 1962* (Paris: PUF, 1964).

Sergent, Edmond and L. Perrot, *La Découverte de Laveran – Constantine, 6 novembre 1880* (Paris: Masson, 1929).

Serre, Louis, *Les Arabes Martyres: Étude sur l'insurrection de 1871 en Algérie* (Paris: E. Lachaud, 1873).

Shorter, Aylward, *Cross and Flag in Africa: The 'White Fathers' during the Colonial Scramble* (Maryknoll: Orbis, 2006).

Silverstein, Paul A. and Ussama Makdisi, 'Introduction: Memory and Violence in the Middle East and North Africa', in Makdisi Ussama, and Paul A. Silverstein (eds), *Memory and Violence in the Middle East and North Africa* (Bloomington: Indiana University Press, 2006), 1–24.

Simon, Frédéric, *Les Spahis et les Smalas* (Constantine: L. Marle, 1871).

Simpson Fletcher, Yaël, 'The Politics of Solidarity: Radical French and Algerian Journalists and the 1954 Orléansville Earthquake', in Lorcin, Patricia M.E. (ed.), *Algeria and France 1800–2000: Identity, Memory, Nostalgia* (Syracuse: Syracuse University Press, 2006). 84–98.

Siouti, Sidi, *Livre de la Miséricorde dans l'art de guérir les maladies et de conserver la santé*, with an introduction by E. Bertherand (Paris: J.B. Baillière, 1856).

Six, Jean-François, Maurice Serpette and Pierre Sourisseau, *Le testament de Charles de Foucauld* (Paris: Fayard, 2005).

Smith, Barbara Leigh, *Women and Work* (London: Bosworth and Harrison, 1857).

Smith, David M., *Moral Geographies: Ethics in a World of Difference* (Edinburgh: Edinburgh University Press, 2000).

Snedden, Robert, *Medical Ethics: Changing Attitudes 1900–2000* (Hove: Wayland, 1999).

Spillmann, Georges, *Napoléon III et le royaume arabe d'Algérie* (Paris: Académie des sciences d'outre-mer, 1975).

Stora, Benjamin, 'The Algerian War in French Memory: Vengeful Memory's Violence', in Makdisi, Ussama and Paul A. Silverstein (eds), *Memory and Violence in the Middle East and North Africa* (Bloomington: Indiana University Press, 2006), 151–174.

Strachan, John, 'The Pasteurization of Algeria?' *French History*, 20/3 (2006), 260–275.

Sumner, Charles, *White Slavery in the Barbary States* (Boston: William D. Ticknor, 1847).

Tableau de tous les traitements et salaires payés par l'état d'après le budget de 1830, par un membre de la société de statistique de France (Paris: Hautecoeur-Martinet, 1831).

Taithe, Bertrand, *Defeated Flesh: Welfare, Warfare and the Making of Modern France* (Manchester: Manchester University Press, 1999).

Taithe, Bertrand, 'The Red Cross Flag in the Franco-Prussian War: Civilians, Humanitarians and War in the "Modern Age"', in Cooter, Roger, Mark Harrison and Steve Sturdy (eds), *War, Medicine and Modernity* (Stroud: Sutton, 1998), 22–47.

De Tchihatchef, P., *Espagne, Algérie et Tunisie: Lettres à Michel Chevalier* (Paris: J.B. Baillière, 1880).

Temime, Emile, *Un Rêve méditerranéen: Des saint-simoniens aux intellectuels des années trente (1832–1962)* (Arles: Actes Sud, 2002).

Temimi, Abdeljelil, *Le Beylik de Constantine et Hadj 'Ahmed Bey' (1830–1837)* (Tunis: Revue d'Histoire Maghrébine, 1978).

Temimi, Abdeljelil, *Recherches et Documents d'Histoire Maghrébine: L'Algérie, la Tunisie et la Tripolitaine (1816–1871)* (Tunis: Revue d'Histoire Maghrébine, 1980).

Tessler, Mark, *Area Studies and Social Science: Strategies for Understanding Middle East Politics* (Bloomington: Indiana University Press, 1999).

Thobie, Jacques, Gilbert Meynier, Catherine Coquery-Vidrovitch and Charles-Robert Ageron, *Histoire de la France coloniale: 1914–1990* (Paris: Armand Colin, 1990).

Thompson, Victoria, '"I Went Pale with Pleasure": The Body, Sexuality and National Identity among French Travelers to Algiers in the C19', in Lorcin, Patricia M.E. (ed.), *Algeria and France 1800–2000: Identity, Memory, Nostalgia* (Syracuse: Syracuse University Press, 2006). 18–32.

Thomson, David, *France: Empire and Republic, 1850–1940, Historical Documents* (London: Palgrave Macmillan, 1968).

de Tocqueville, Édouard, *Des enfants trouvés et des orphelins pauvres comme moyen de colonisation de l'Algérie* (Paris: Aymot, 1850).

Todorov, Tzvetan, *The Conquest of America: The Question of the Other* (Oklahoma: Oklahoma University Press, 1993).

Toulmin, Stephen, 'How Medicine Saved the Life of Ethics', *Perspectives on Biology and Medicine*, 25 (1980–81), 736–750.

Trapani, D.G., *Alger tel qu'il est ou tableau statistique, moral et politique de cette régence* (Paris: L. Fayolle, 1830).

Trottier, François *Boisement et colonisation: Note sur l'eucalyptus et subsidiairement sur la nécessité du reboisement de l'Algérie* (Algiers: F. Paysant, 1876).

Trumelet, Corneille, 'Notes pour servir à l'histoire de l'insurrection dans le sud de la province d'Alger de 1861 à 1869 and 1887, *Revue Africaine*, journal publié par la Société historique d'Alger (1883), 36–75.

Turin, Yvonne, *Affrontements culturels dans l'Algérie coloniale: écoles, médecines, religion, 1830–1880* (Paris: François Maspéro, 1971).

Ullmann, M., *Islamic Medicine* (Edinburgh: Edinburgh University Press, 1978).

Un ancien capitaine de Zouaves, *Les Grottes de Dahara: Récit historique* (Paris: M. Blot. 1864).

Un ancien officier de l'armée d'Afrique, *L'Algérie devant l'Assemblée Nationale: causes des insurrections algériennes* (Versailles: Muzard, 1871).

Un chef de Bureau Arabe, *L'Algérie Assimilée: Étude sur la constitution et la réorganisation de l'Algérie* (Constantine: L. Marle, 1871).

Un Colon, *Quelques arguments en faveur de la colonisation européenne en Algérie* (Algiers: A. Bouyer, 1863).

Unschuld, Paul U., *Medical Ethics in Imperial China: A Study in Historical Anthropology* (Berkeley: University of California Press, 1979).

Urbain, Ismaÿl, *L'Algérie pour les algériens*, ed. Levallois, Michel (Paris: Séguier, 2000).

de Valmy, M. le Duc, *Question d'Alger: histoire des négotiations* (Paris: Proux, 1845).

Van Vollenhoven, Joost, *Essai sur le fellah algérien* (Paris: Arthur Rousseau, 1903).

Vann, Michael G. 'The Good, the Bad and the Ugly: Variation and Difference in French Racism in Colonial Indochine', in Peabody, Sue and Tyler Stovall (eds), *The Color of Liberty: Histories of Race in France* (Durham: Duke University Press, 2003), 187–205.

Variot, Joseph, *Les Pères Blancs ou Missionaires d'Alger* (Lille: Société St Augustin, 1887).

Vaughan, Megan, *Curing Their Ills: Colonial Power and African Illness* (Cambridge: Polity Press, 1991).

Veatch, Robert M., *Disrupted Dialogue: Medical Ethics and the Collapse of Physician-Humanist Communication (1770–1980)* (Oxford: Oxford University Press, 2005).

Védrènes, A., 'De la trépanation du crâne chez les indigènes de l'Aurès (Algérie)', *Revue de Chirurgie*, 5 (1885).

Vereker, C.S., *Scenes in the Sunny South, Including the Atlas Mountains and the Oases of the Sahara in Algeria*, 2 vols (London: Longmans, 1871).

Vialar, M. le Baron, *Simples faits exposés à la Réunion Algérienne du 14 avril 1835* (Paris: Firmin Didot).

Vidal-Bué, Marion, *Alger et ses peintres 1830–1960* (Paris: Paris-Mediterranée, 2000).

Vidal-Bué, Marion, *Les peintres de l'autre rive: Alger 1830–1930* (Cannes: Musée de la Castres, 2003).

Vidal-Nacquet, Pierre (ed.), *Les crimes de l'armée française* (Paris: François Maspéro, 1975).

Vidal-Nacquet, Pierre, *La Torture dans la République (1954–1962)* (Paris: Les Editions de Minuit, 1988).

Vidal-Nacquet, Pierre, *L'Affaire Audin (1957–1978)* (Paris: Les Editions de Minuit, 1989).

Villot, Le Capitaine, *Mœurs, coutûmes et institutions des indigènes de l'Algérie,* (Constantine: L. Arnolet, 1871).

Viollier, Georges, *Les Deux Algérie* (Paris: Librairie Paul Dupont, 1898).

Volland, M. Le baron, *Réfutation du rapport de la commission du budget, en ce qui concerne nos possessions en Afrique* (Paris: L.E. Herhan, 1835).

Waltraud, Ernst and Bernard Harris (eds), *Race, Science and Medicine, 1700–1900* (London: Routledge, 1999).

Warnery, A., *Résumé de la situation morale et matérielle de l'algérie* (Paris: E. Brière, 1847).

Warnier, Dr A., *L'Algérie devant le sénat* (Paris: Challamel Ainé, 1863).

Warnier, Dr A., *L'Algérie devant l'opinion publique: indigènes et immigrants, examination rétrospectif* (Paris: Challamel Ainé, 1864).

Watson, William E., *Tricolor and Crescent: France and the Islamic World* (Westport: Greenwood, 2003).

Watts, Sheldon, *Epidemics and History: Disease, Power and Imperialism* (New Haven: Yale University Press, 1997).

Wear, Andrew, Joanna Geyer-Kordesch and Roger French (eds), *Doctors and Ethics: The Earlier Historical Setting of Professional Ethics* (Amsterdam: Rodopi, 1993).

Weil, Patrick, 'Le statut de musulmans en Algérie, une nationalité française dénaturée' in Banat-Berger, Françoise (ed.), *La Justice en Algérie 1830–1962* (Paris: La documentation Française, 2005), 95–109.

Wood, Nancy, *Vectors of Memory: Legacies of Trauma in Postwar Europe* (Oxford: Berg, 1999).

Worboys, Michael, 'Reviews', *Social History of Medicine*, 8/2 (1995), 331–334.

Wright, Gwendolyn, *The Politics of Design in French Colonial Urbanism* (Chicago: University of Chicago Press, 1991).

Wright, Henry C., *Defensive War Proved to be a Denial of Christianity and of the Government of God* (London: Charles Gilpin, 1846).

Wyschogrod, Edith, *An Ethics of Remembering: History, Heterology and the Nameless Others* (Chicago: University of Chicago Press, 1998).

Yacoub, Ahmed Abdel Aziz, *The Fiqh of Medicine: Responses in Islamic Jurisprudence to Developments in Medical Science* (London: Ta-Ha, 2001).

Young, Robert J.C. *White Mythologies: Writing History and the West* (London: Routledge, 1990).

Young, Robert J.C., *Postcolonialism: An Historical Introduction* (Oxford: Blackwell, 2001).

Zinebe, *Nouvelle Kabyle: Souvenirs de l'Insurrection de 1871 dans la région de Dellys racontés par un vieux colon* (Algiers: F. Montégut, 1918).

Websites

http://www.saphirnews.com/Docteur-Philippe-Grenier_a1740.html – 16.07.06
http://www.le-carrefour-de-lislam.com/Grenier/grenier1.htm – 16.07.06
http://www.marxists.org/history/algeria/1963/09/constitution.htm – 16.07.06

Index